AF341983

HORIZONS IN NEUROSCIENCE RESEARCH

VOLUME 25

HORIZONS IN NEUROSCIENCE RESEARCH

Additional books in this series can be found on Nova's website
under the Series tab.

Additional e-books in this series can be found on Nova's website
under the eBook tab.

HORIZONS IN NEUROSCIENCE RESEARCH

VOLUME 25

ANDRES COSTA

AND

EUGENIO VILLALBA

EDITORS

Nova Biomedical

New York

NOTICE TO THE READER

Library of Congress Cataloging-in-Publication Data

ISBN: 978-1-63485-286-9

ISSN: 2159-113X

Published by Nova Science Publishers, Inc. † New York

CONTENTS

Preface   **vii**

Chapter 1   Science and Living Human Brain: Clinical
Exploration and Manipulation of the Neural
Networks and Structural Lesions   **1**
Kemal Dizdarevic and Samir Delibegovic

Chapter 2   The Role of Theta and Gamma Oscillations
in Working Memory Capacity   **27**
Anja Pahor and Norbert Jaušovec

Chapter 3   The Potential Usage of Plastic Clip
in Cerebrovascular Neurosurgery   **49**
Samir Delibegovic and Kemal Dizdarevic

Chapter 4   Spinal Root Avulsion: Motoneuron Death
and Strategies to Promote Survival
of Injured Motoneurons   **59**
Heng Li, Carolin Ruven and Wutian Wu

Chapter 5   Spatial Patterns of Phosphorylated TDP-43-
Immunoreactive Cellular Inclusions in Familial
and Sporadic Frontotemporal Lobar Degeneration
with TDP-43 Proteinopathy   **89**
Richard A. Armstrong

Chapter 6   Neuropathological Aspects of Acute
Disseminated Encephalomyelitis   **109**
Shinji Ohara

Chapter 7 "Meningitis-Retention Syndrome" as a Mild Form
of Acute Disseminated Encephalomyelitis   **123**
*Ryuji Sakakibara, Fuyuki Tateno, Masahiko Kishi,
Yohei Tsuyusaki, Yosuke Aiba, Hiromi Tateno,
Tsuyoshi Ogata, Tatsuya Yamamoto,
Tomoyuki Uchiyama and Tomonori Yamanishi*

Chapter 8 Pure Motor Monoparesis Due to Ischemic Stroke   **167**
Akiyuki Hiraga

Chapter 9 Stroke Associated with Nephrotic Syndrome -
Acute Ischemic Stroke, Cerebral Venous Sinus
Thrombosis, and Intracerebral Hemorrhage   **197**
Masaru Kuriyama

Index   **225**

PREFACE

This book provides readers with the latest developments in neuroscience research. Chapter One discusses clinical exploration and manipulation of the neural networks and structural lesions of the human brain. Chapter Two studies the role of theta and gamma oscillations in working memory capacity. Chapter Three examines the potential usage of plastic clips in cerebrovascular neurosurgery. Chapter Four reviews avulsion-induced motoneuron death and discusses different strategies to facilitate survival of motoneurons following spinal root avulsion. Chapter Five discusses spatial patterns of phosphorylated TDP-43-immunoreactive cellular inclusions in familial and sporadic frontotemporal labor degeneration with TDP-43 proteinopathy. Chapter Six reviews neuropathological aspects of acute disseminated encephalomyelitis. Chapter Seven identifies the frequency, clinical symptoms, urodynamic findings, putrative underlying pathology, and management of meningitis–retention syndrome (MRS). Chapter Eight provides an update on clinical features, lesion topography, and prognosis of pure motor monoparesis (PMM) due to ischemic stroke. The final chapter discusses strokes associated with nephrotic syndrome.

Chapter 1 - Human brain, as the most complex piece of matter in our known universe and structural base for our mind, is functioning through interactions of its networks of neurons connected in a tidy grid that resembles a city street map. These specific connections between neurons make the connectome responsible for our unique identity and are ruled by principles that can be discovered by scientific methods. Clinical neuroscientists attempt to deduce the functions of the brain from the behavioral effects of brain damage or to assess the different effects of new type of treatment. Basic neuroscientists developed the techniques for observing the brain in controlled experiments permitting them to address testable hypothesis about distinct aspects of brain functioning. The fundamental search of neuroscientists are directed to localizing of mental processes and understanding the mind through specific brain region activity. The USA Congress designated the 1990s the 'Decade of the Brain'. Conceptual advances during the decade of the brain were important and include: a) stem cells in the nervous system; b) brain plasticity: c) mito-chondrial disease; d) genetic basis of neurological disease and neo-plasms; e) imaging of the central nervous system; f) neuroactive growth factors, peptides and cytokines. In the beginning of the 21st century some failures persist including treatment of central nervous system injury, transplantation and regeneration in CNS as well as stroke therapy espe-cially the brain protection from ischemia and mind-brain relationships. Neurosurgeons are the only neuroscientists who work with the human living brain and the anatomical and physiological substrate of the human mind. So neurosurgeons should play a pivotal role in directing the advances in neuroscience. Our ignorance of brain is still so vast that the "decade of the brain" should be re-designated into the "century of the brain" implying that authors' need a whole new century to understand major brain functions. Through exploring the human brain, clinically or expe-rimentally, the picture of our inner universe becomes more connected but not absolutely exact. Obviously, authors' create the chance for us only thanks to our brain.

Chapter 2 - Two experiments were conducted in order to investigate the relationship between theta and gamma frequencies, and the ratio between these frequencies, and working memory. In experiment 1, participants' resting EEG data was recorded after which they solved verbal and spatial tests of short term memory (STM) capacity. As predicted, theta/gamma cycle length ratio positively correlated with verbal STM capacity. In contrast, theta/gamma cycle length ratio positively correlated with the forward version of a spatial STM test and negatively correlated with the backward version of the test. Individual gamma frequency positively correlated with verbal and spatial STM capacities, whereas individual theta frequency only correlated with performance on the spatial STM tests. In order to verify the results of experiment 1, a second experiment was conducted in which either theta or gamma transcranial alternating current stimulation (tACS) was applied over parietal areas and contrasted against sham stimulation. Following sham or verum tACS, the participants solved a series of figural and verbal n-back tasks during which their EEG was recorded. The results showed that theta tACS significantly reduced the average reaction time on a difficult verbal n-back test, which was accompanied by changes in early visual processing.

Collectively, these results provide further insight into the neural basis of working memory.

Chapter 3 - The fourth generation of clips used in neurosurgery are made of titanium and its alloys. Titanium clips are standard in cerebrovascular operative neurosurgery. Apart from this, plastic clips are not applied in neurosurgery at all, although they have some practical advantages. One well-known advantage is connected with their intrinsic diamagnetic and nonconductive characteristic, that results in fewer CT and MRI artifacts than titanium clips. The small difference in the inflammatory reaction between the titanium clip and the plastic clip in the animal canine model demonstrates the possibility of the potential use of plastic clips in operative neurosurgery. Also a three-dimensional 'finite element model' for a Hem-o-lok clip, placed on the aneurysmal neck of an imaginary intracranial supraclinoid internal carotid artery, and analysis of the influence of blood pressure on the biomechanical characteristics of the Hem-olok clip showed that Hem-o-lok plastic meets the necessary criteria for intracranial application and aneurysmal clipping procedures. At a maximum blood pressure of 250 mm Hg, the clip did not open at all causing blood leakage in the perivascular space. Also, the output data did not show any tension above the clip material's breakdown limit. Many characteristics of plastic clips, including weight, strength of closing, biocompatibility and low price, make them attractive and potentially suitable for usage in cerebrovascular neurosurgery. Typically, clips are repositioned several times during microsurgical clipping procedures, before their definite placement on the aneurysmal neck. Therefore, it is vital that they are easily placed and removed. Unfortunately, Hem-o-lok clips need another instrument for removal, which is a time-wasting procedure. It remains the task of the manufacturers to facilitate the application of these clips in the limited intracranial space. The potential usage of the plastic clips described in this study could open new clinical perspectives.

Chapter 4 - Spinal root avulsion is common in brachial plexus injury, with the rupture site located at the CNS/PNS transitional zone close to the motoneuron cell body. Progressive and rampant motoneuron death follows with a root avulsion, accompanied by profound changes of motonuerons in both cellular morphology and biochemical profile. Functional restoration firstly requires survival of the injured motoneurons. Different strategies have been reported to rescue the lesioned motoneurons from death, which include: 1) Neurotrophic factors such as glial cell line-derived neurotrophic factor (GDNF) and brain-derived neurotrophic factor (BDNF) are efficient in promoting the survival of motoneurons. 2) Surgical repair, including implantation of peripheral nerve graft or reimplantation of the avulsed roots to the ventral spinal cord, could not only reduce motoneuron degeneration in both acute and delayed injury model, but also allow axons to regenerate into the nerve conduit. 3) With the coincidence of avulsion-induced cell death and the up-expression of nitric oxide synthase (NOS), manipulation of NOS is able to decrease motoneuron death. 4) As apoptosis has been indicated as one of the mechanisms leading to motoneuron death, inhibition of the caspase cascade promoted motoneuron survival. 5) In addition, successful axonal regeneration into the peripheral nerve trunk might be beneficial to the lesioned motoneurons. This chapter will review the avulsion-induced motoneuron death and discuss different strategies to facilitate survival of motoneurons following spinal root avulsion.

Chapter 5 - Abnormal protein aggregates of transactive response (TAR) DNA-binding protein (TDP-43) in the form of neuronal cytoplasmic inclusions (NCI), oligodendroglial inclusions (GI), neuronal internuclear inclusions (NII), and dystrophic neurites (DN) are the pathological hallmark of frontotemporal lobar degeneration with TDP-43 proteinopathy (FTLD-TDP). To investigate the role of phosphorylated TDP-43 (pTDP-43) in neurodegeneration in FTLD-TDP, the spatial patterns of the pTDP-43-immunoreactive NCI, GI, NII, and DN were studied in frontal and temporal cortex in three groups of cases: (1) familial FTLD-TDP caused by *progranulin* (*GRN*) mutation, (2) a miscellaneous group of familial cases containing cases caused by *valosin-containing protein* (VCP) mutation, *ubiquitin associated protein 1 (UBAP1)* mutation, and cases not associated with currently known genes, and (3) sporadic FTLD-TDP. In a significant number of brain regions, the pTDP-43-immunoreactive inclusions developed in clusters and the clusters were distributed regularly parallel to the tissue boundary. The spatial patterns of the inclusions were similar to those revealed by a phosphorylation-independent anti-TDP-43 antibody. The spatial patterns and cluster sizes of the pTDP-43-immunoreactive inclusions were similar in *GRN* mutation cases, remaining familial cases, and in sporadic FTLD-TDP. Hence, pathological changes initiated by different genetic factors in familial cases and by unknown causes in sporadic FTLD-TDP appear to follow a parallel course resulting in very similar patterns of degeneration of frontal and temporal lobes.

Chapter 6 - Acute disseminated encephalomyelitis (ADEM) has been generally considered as an immunologically-mediated inflammatory demyelinating disease of the CNS. Unlike Multiple Sclerosis (MS), which is characterized by clinical relapses and progression over years, ADEM is usually monophasic and relatively with a good prognosis. Although, ADEM and MS have been long considered as separate disease entities, clinical differentiation of ADEM from the first attack of MS is often difficult because of overlapping clinical features. Pathologically, perivenous demyelination and discrete confluent demyelination have been generally regarded as the hallmark of ADEM and MS, respectively. However, hybrid cases showing cardinal pathological features of both ADEM and acute stages of MS do exist, suggesting that ADEM may share some common underlying pathologic mechanisms with certain stages or subgroups of MS. Some patients with clinically or pathologically diagnosed ADEM may later develop MS, although others apparently do not. Factors that may be responsible for such differences are currently unknown, although the presence of certain acquired mechanisms to suppress later disease development may be a possibility. Moreover, cases of ADEM with concomitant peripheral nerve involvement have been documented clinically and pathologically, which may suggest common features underlying immune etiologies involving both CNS and PNS.

Chapter 7 - *Aims*: A peculiar combination of acute urinary retention and aseptic meningitis has been described. This combination is referred to as meningitis–retention syndrome (MRS), since patients with this syndrome exhibited no other abnormalities, except for mild pyramidal involvement. The authors aimed to delineate this syndrome by reviewing literatures. *Methods*: The authors performed a systematic review of the literature to identify the frequency, clinical symptoms, urodynamic findings, putrative underlying pathology, and management of this syndrome. *Results*: Patients with MRS have typical symptoms of fever, headache, stiff neck, and minor pyramidal signs, together with acute urinary retention. The bladder is initially areflexic, but soon becomes either normal or overactive in the repeated urodynamics during the course of the disorder. MRS is thought to be a very mild form of acute disseminated encephalomyelopathy (ADEM), with increased cell count, total protein, and occasional myelin basic protein in the cerebrospinal fluid. Proper management of the acute urinary retention is necessary to avoid bladder injury due to overdistension. The effectiveness of immune treatments (e.g., steroid pulse therapy) in shortening the urinary retention period awaits further study. *Conclusions*: Although rare, MRS is thought to be a very mild form of ADEM, and a disorder that both urologists and neurologists may encounter. MRS should be listed in the differential diagnosis of acute urinary retention.

Chapter 8 - Pure motor monoparesis (PMM), an isolated paresis due to ischemic stroke in a single arm or leg without higher, cranial, or sensory dysfunction, is an uncommon clinical condition. Cortical infarctions of the precentral knob and anterior cerebral artery territory are the most commonly reported PMM lesions affecting the upper and lower extremities, respectively. Other reported lesion sites include the subcortex, corona radiata, internal capsule, and brainstem; however, they are reported less frequently than cortical infarctions because the descending motor axons in these areas are highly compacted. Large artery atherosclerosis and cardiogenic embolism are considered as the two main etiologies of cortical infarction. In earlier studies, PMM cases due to lacunar infarctions were presumed to be rare. However, lacunar infarction PMM cases have been increasingly reported in the recent years. The weakness patterns in PMM are complex and include proportional and distal- or proximal-dominant weakness. Distal-dominant weakness is the most common PMM type. Some patients exhibit isolated finger paresis or predominant weakness of specific fingers. Although the overall prognosis is favorable in PMM due to ischemic stroke, short- and long-term prognosis depends on risk factors and stroke mechanisms. The diagnosis is sometimes difficult because many cases of PMM lack pyramidal signs and mimic peripheral nerve damage. Therefore, acute monoparesis, particularly in elderly patients with conventional risk factors, should be carefully assessed. Diffusion-weighted imaging is the most useful diagnostic tool for PMM due to ischemic stroke as small cortical infarctions often cannot be detected by conventional sequences of magnetic resonance imaging, or computed tomography. This review provides an update on clinical features, lesion topography, and prognosis of PMM due to ischemic stroke.

Chapter 9 - Nephrotic syndrome is defined by the presence of heavy proteinuria, hypoalbuminemia, hyperlipidemia and peripheral edema. Thrombotic diseases are also frequently observed. Arterial and venous thromboses are potential complications of nephrotic syndrome. Arterial thromboses are less frequent than venous thromboses and the most common locations are femoral arteries, although other arteries may be involved. Stroke associated with nephrotic syndrome has been reported in a number of case reports, but it is rare and its clinical details remain obscure. The authors identified ten cases of acute ischemic stroke, two cases of cerebral venous sinus thrombosis, and four cases of intracerebral hemorrhage by screening the hospitalized patients enrolled in their stroke unit. The incidence of acute ischemic stroke associated with nephrotic syndrome was 0.09% of all kinds of stroke and 0.12% of acute ischemic stroke, and was fivefold that of cerebral venous sinus thrombosis. The subtypes of stroke identified in the authors' screening were large-artery atherosclerosis (six instances), small-vessel occlusion (three instances), and cardioembolism (one instance). The results of a retrospective cohort study showed that acute ischemic stroke-associated nephrotic syndrome is strongly associated with atherosclerosis of the cerebral artery, especially in the anterior circulation. Patients also exhibited atherosclerosis of the internal carotid and lower extremity arteries. The causal disorder of nephrotic syndrome was diabetic nephropathy in eight of these cases. Twenty-five patients with cerebral venous sinus thrombosis were admitted to our hospital during the same period. Nephrotic syndrome has been reported to be responsible for 0.6-1.0% of all cases of cerebral venous sinus thrombosis; however, the authors' study showed that this percentage may be higher. Nephrotic syndrome essentially induced a hypercoagulable state. However, four cases of intracerebral hemorrhage associated with nephrotic syndrome were identified by the screening. These exhibited the presence of strong risk factors for hemorrhage, including hypertension, vascular vulnerability with atherosclerosis or amyloidosis, and change of bloodstream after operation, suggesting that they overcame the risk for thrombophilia. The diseases associated with nephrotic syndrome were diabetic nephropathy and amyloidosis in three cases and one case, respectively. Nephrotic syndrome tends to be associated with a risk for arterial or venous thrombosis. In addition, the authors must pay attention to intracerebral hemorrhage associated with nephrotic syndrome in cases of stroke.

In: Horizons in Neuroscience Research. Vol. 25   ISBN: 978-1-63485-286-9
Editors: A. Costa and E. Villalba   © 2016 Nova Science Publishers, Inc.

Chapter 1

SCIENCE AND LIVING HUMAN BRAIN: CLINICAL EXPLORATION AND MANIPULATION OF THE NEURAL NETWORKS AND STRUCTURAL LESIONS

Kemal Dizdarevic[1], MD, PhD
and Samir Delibegovic[2], MD, PhD
[1]Department of Neurosurgery, University Clinical Center,
Faculty of Medicine, University of Sarajevo,
Bosnia and Herzegovina
[2]Department of Surgery, University Clinical Center,
Faculty of Medicine, University of Tuzla,
Bosnia and Herzegovina

ABSTRACT

Human brain, as the most complex piece of matter in our known universe and structural base for our mind, is functioning through interactions of its networks of neurons connected in a tidy grid that resembles a city street map. These specific connections between neurons make the connectome responsible for our unique identity and are ruled by principles that can be discovered by scientific methods. Clinical neuroscientists attempt to deduce the functions of the brain from the behavioral effects of brain damage or to assess the different effects of new type of treatment. Basic neuroscientists developed the techniques for observing the brain in controlled experiments permitting them to address testable hypothesis about distinct aspects of brain functioning. The fundamental search of neuroscientists are directed to localizing of mental processes and understanding the mind through specific brain region activity.

The USA Congress designated the 1990s the 'Decade of the Brain'. Conceptual advances during the decade of the brain were important and include: a) stem cells in the nervous system; b) brain plasticity: c) mitochondrial disease; d) genetic basis of neurological disease and neoplasms; e) imaging of the central nervous system; f) neuroactive growth factors, peptides and cytokines. In the beginning of the 21st century some failures persist including treatment of central nervous system injury, transplantation and regeneration in CNS as well as stroke therapy especially the brain protection from ischemia and mind-brain relationships. Neurosurgeons are the only neuroscientists who work with the human living brain and the anatomical and physiological substrate of the human mind. So neurosurgeons should play a pivotal role in directing the advances in neuroscience. Our ignorance of brain is still so vast that the "decade of the brain" should be re-designated into the "century of the brain" implying that we need a whole new century to understand major brain functions. Through exploring the human brain, clinically or experimentally, the picture of our inner universe becomes more connected but not absolutely exact. Obviously, we create the chance for us only thanks to our brain.

INTRODUCTION

Human brain is a heterogenous, multi-dimensional and multi-functional organ with phylogenic, ontogenic, histogenic and immunogenic specific regions and different morphological and functional compartments. The whole brain is eloquent, not just its certain region. The brain is more than simple sum of its parts. The living human brain is functioning through interactions of its parts. The connectivity within the brain is more important than its particular centers [1, 2].

Certainly, human brain is the most complex piece of matter in our known universe but ironically and intentionally closed in very simple and rigid container of the skull.

CONNECTOME

What is it that makes us ? It can be said that we are our connectome. The connectome refers to the interconnected network of neurons. The effort to map the connectome and decipher the electrical signals that zap through it to generate our thoughts, feelings, and behaviors has become possible through development of powerful new tools and technologies.

New advances in computer science, math, imaging and data visualization are empowering us to study the human brain as an entire organ, and at a level of detail not previously imagined possible in a living person.

A typical human brain contains 100 billion neurons, each with about 10,000 connections. The connectome will also give us a new tool to explore how genes influence the brain's connections - and how behavioral and environmental factors act to sculpt those connections.

A recent study by connectome researchers, published in the journal Science, revealed that the brain's neurons are not the haphazard tangle, but they are arranged in a tidy grid that resembles a city street map. We might think that the cells of our brain form connections in random places and angles. But our mental circuitry consists of sheets of fibres that intersect at right angles, with no diagonals anywhere to be seen.

Our neurons are surrounded by insulation (myelin), which shields the electric currents that run down their length. But this insulation is a recent evolutionary innovation. In some earlier brains, which don't have myelin, neurons would have electrical problems if they crossed at anything other than a right angle. They would be more likely interfere with each other [1].

SCIENCE

Even the most complex systems (such as the human brains) operate according to simple laws accessible to the average person.

Everything around us are ruled by principles that can be discovered by scientific methods.

Observing the world is the first step toward science. Experiment controls the system being studied changing only one thing at a time to see what happens. For example, a cancer research will keep all elements of the treatment of tumors in experimental animals the same except for one. In experiment, scientist wants to make a detailed examination of one piece in isolation. Scientist is supposed to be open-minded with ability to ignore preconceptions and follow the fact whenever they lead. Observation or experiment are both integral part of scientific method [3, 4].

The central idea of science is that we can learn about the laws that govern natural events by experimentation and then produce predictive theories. We live in ordered universe whose working are accessible to the human mind (brain). Science unlike religion does not seek absolute truth, but instead uses a method that produce better understanding of reality. This method begins with a question: Why do things happen this way and not some other way? [3]

Science is based on evidence and uses arguments and facts. On the order hand, philosophy do not care about the evidence, but it also uses arguments and facts.

The good scientific theory will make testable predictions. We can go back to experiment or observation to see whether these predictions are valuable. In science, there are right and exact answers. In philosophy or literature (the humanities) there is no objective arbiter that plays the role of nature. If an idea can't be tested against experiment, if it can't be made to confront nature, it is not science! This means that scientific ideas must produce statements that can be tested.

Science is falsifiable: it must be possible to imagine an experimental or observational result that would show the theory to be wrong.

Scientific cycle is to do experiment, to discover regularities, to creat theory, to make predication, then the return to experiment to test the prediction. In science, there is always room for new ideas and extending the knowledge

NEUROSCIENTIST

The main goal of neuroscientist is to understand how the brain and whole nervous system function. Neuroscientists may be divided into two groups: clinical and basic (experimental).

Clinical neuroscientists are physicians (medical doctor) working in medical disciplines associated with the human nervous system (neurosurgery, neurology, psychiatry and neuropathology). They are attempting to deduce the functions of various parts of the brain from the behavioral effects of brain damage. Others conduct clinical studies and researches to assess the benefits and risks of new type of treatment. But the foundation for majority of medical treatments of the brain(nervous system) continues to be laid in research performed by experimental neuroscientists (neuroanatomists, neurophysiologists, molecular neurobiologists, computational neuroscientists, neuropsychologists etc) [5, 6].

BASIC NEUROSCIENCE

Basic Neural science (Basic Neuroscience) is the modern science of the brain which emerged in the mid 1970s with the convergence of molecular biology, cell biology, neuroanatomy, and neurophysiology. Development of techniques for observing the brain in controlled experiments permits the neuroscience to address testable hypothesis about different aspect of brain functioning.

Several disciplines including cognitive neuroscience, psychology, linguistics, information theory tried to explore how we think. Cognitive neuroscience researches our mental functioning in the context of cell and molecular biology.

Neuroscientist work according to the scientific process which consists of four steps: observation, replication, interpretation and verification.

Observations are made during experiments designed to test a hypothesis. Other types of observation derive from watching the world around us, or from introspection, or from clinical cases. Replication of observation that means repeating the experiment on different subjects as many times as necessary to rule out the possibility that the observation occurred by chance. Interpretation depends on the state of knowledge at the time the observation was made and neuroscientist's preconceived notions. Important step forward are sometimes made when old observations are reinterpreted in a new light. Verification means that observation is sufficient that it can be recognized by any competent scientist who precisely follows the original protocols. Successful verifications imply that observation is accepted as fact.

THE BRAIN AND BEHAVIOR

All behavior is the result of brain function. All disorders which influence behavior are based on some kind of brain disturbances such as in the case of psychiatric (affective and cognitive) or neurologic/neurosurgical (structural) disorders [5, 7, 8]. Neuroscience explains human behavior in terms of the activities of the brain. When we say behavior, we do not think only about simple motor behavior (walking, eating) but all the complex cognitive actions (thinking, speaking- language, and creative works- art).

There are two fundamental questions which are confronted by neuroscience: 1) Are particular mental processes localized to specific regions of the brain; 2) Does the mind represent activity of the whole brain? [5]

EVOLUTION AND STRUCTURE

In vertebrates the nervous system starts as a midline groove on the back of the embryo. This groove forms a thick-walled tube. The front end of the tube which is closed, swells into three connected vesicles which will form the forebrain (prosencephalon), the midbrain (mesencephalon), and the hindbrain. Forebrain divides into an expanded endbrain (telencephalon) and a between-brain (diencephalon) [9, 10].

Evolution of the brain can be understood by comparing the living brains of fish, reptiles and mammals [10]. The progressive enlargement of the endbrain is striking thing in the course of evolution. In all three cases it develops to form two cerebral hemispheres that are connected by very large bundle of axons (the corpus callosum). In all mammals the surface of the cerebral hemispheres is a sheet of cells (neocortex) [9, 10, 11]. The human brain has a neocortex 2-5 mm thick (the grey matter) that covers masses of axons (the white matter). Deep within the white matter of each hemisphere are three collections of neurons: the basal ganglia, the hippocampus (in cross-section looks like a "sea-horse") and amygdala (almond). The basal ganglia are involved in movement control, hippocampus and amygdala have important role in memory and emotion. The thalamus (inner room) is a part of diencephalon and some kind of great relay station with connections in both directions [12, 13, 14].

In mammals, the roof of the midbrain swells into four domes. One pair is mainly concerned with eye movements (colliculi superiores). The other pair is important for auditory information (colliculi inferiores) [11].

Brain stem includes the midbrain (mesencephalon), the pons and the medulla. Tegmentum as a dorsal part of brain stem has the reticular formation that includes nuclei divided according to the neurotransmitters in their neurons. The part of the hindbrain is expanded into highly folded structure which is concerned with balance and posture, fine movement control, and importantly with learning of complex patterns of movement (the cerebellum).

The microneurosurgeon have stored the necessary instructions for their manual work in their cerebellum. In all vertebrates, caudal part of hindbrain (the medulla) is concerned with the control of the circulation and respiration.

Limbic areas called allocortex displey features that range from a pattern of no lamination (septal regions, substantia innominata, amygdala) to two-layered cortex (hippocampal and piriform cortex). The mesocortex includes paralimbic structures (the caudal orbitofrontal cortex, insula, temporal pole, para-hippocampal cortex, and cingulate gyri). Mesocortical structures are interposed anatomically and connectively between allocortex and neocortex. These structures have three to five layers, depending on their proximity to either allocortex or neocortex.

Systems in the human brain that are responsible for sensations, perceptions, voluntary movement, learning, language, language, and cognition all converge in the neocortex [5].

Cerebral cortex in all vertebrate animals has several common features: a) cortical neuron bodies are arranged in layers; b) the most superficial layer is separated from the surface by a zone that lacks neurons (molecular layer); c) at least on cell layer contains pyramidal cells that emit large dendrites up to molecular layer

Hippocampus (from the Greek- "seahorse") is a link between the cingulate cortex and hypothalamus. Cingulate cortex affects hypothalamus via hippocampus and fornix and it is important part of Papez emotional circuit. Hypothalamus governs the behavioral expression of emotions and it influences cingulate cortex through anterior thalamus [14, 15].

The neocortex is found only in mammals. In fact, over the course of human evolution the neocortex has expended not whole cerebral cortex. The thalamus is the gateway to the neocortex not to the cerebral cortex. The neocortex is divided into zones by Korbinian Brodmann (German neuro-anatomist, at the beginning of the twentieth century). He constructed a cytoarchitectural map of the neocortex [16].

Amount of cortex has changed over the course of evolution but its basic structure has not. The primordial neocortex consisted mainly of three types of cortex: primary sensory areas, secondary sensory areas, motor areas. When we speak of the expansion of the mammalian cortex over the course of evolution we mean that the regions between these areas have expanded. In the human brain a considerable number of areas remains particulary in the frontal and temporal lobes. These areas are the association areas of neocortex. Association neocortex is a more recently developed. It is characteristic of the primate brain. The emergence of the mind that implies our unobservable mental states such as desires, intentions, anticipations, beliefs correlates with the expansion of the frontal cortex [9, 14, 17, 18, 19].

NEANDERTHALS AND PREFRONTAL CORTEX

When we were born, our brains were more elongated, it rounded out into its globular shape as we grew up. Neanderthals were born with brains in that same elongated shape. But in their case brain never changed and adult Neanderthals' brain did not move to the more globular shape like ours. Neanderthals were keeping the same basic shape of their brains throughout life, maintain the pattern of brain development seen in chimpanzees [20].

The Neanderthals had brain the same size as us but our cognitive abilities are significantly higher. The explanations for this fact is internal organization of the neural networks in the human developing brain which include complex connections between diverse brain regions [14, 20].

Although only a few genes separate modern humans from Neanderthals, during evolution we had important competitive advantage over Neanderthals. The main reason for this advantage is our huge prefrontal cortex. It gives globular shape in adult human brain and more importantly, it is morphological basis for our ability to anticipate and to travel in time using our cognition [20, 21, 22].

The prefrontal cortex is responsible for anticipation and working memory which means the ability to hold in mind the facts essential for completing a task or resolving a problem. Working memory is executive function making possible all other intellectual efforts. The prefrontal cortex is also important for self-reflection as a integral part of our conscious [5, 6, 14, 21, 22].

HYPOTHALAMUS

Hypothalamus as another part of diencephalon is almost the center of our inner universe. It controls the pituitary gland which dominates the entire hormonal system in our body (effects on metabolism, growth, reproduction). Hypothalamus acts through the autonomic (vegetative) nervous system that controls events in the body that occur automatically but not necessary unconsciously. Hypothalamus is also the crucial part of the limbic system and it governs the expression of emotions. Additional role of the hypothalamus is to act with other brain's networks in controlling sleep and wakefulness and in producing the physiological changes in the body associated with strong emotions (fear, anger or pleasure) [5, 12, 23].

SEAT OF THE SOUL

Rene Descartes' hypothesis that pineal gland is "seat of the soul" is not true, but hypothalamus can be near of that claim [14, 24]. It converges, directly or indirectly, almost all influences on our life, including those from individual's mind and neocortex. Probably, the pineal gland as a part of epithalamus is the vestige of a third eye, that is found in the fossils of certain fish and reptiles and that still exists in one living reptile (the Tuatara from New Zealand). In human living brain, the pineal secretes a melatonin during dark hours and maintains daily rhythms.

The pineal is stimulated to secrete melatonin through this neuronal network. Interestingly, level of melatonin is higher in the younger age, young brain has more melatonin and is prone to sleep more. Sleep is a readily reversible state of reduced responsiveness to the environment (Coma and general anesthesia are not readily reversible).

AUTONOMIC (VEGETATIVE) NS

The autonomic (vegetative) nervous system is an extensive network of interconnected neurons that are widely distributed inside the body cavity. Via this system brain (hypopthalamus) controls the organism and conducts the mental energy (soul) through the body. The parasympathetic nervous system as a division of autonomic system is concerned with so called "housekeeping functions" (appetite, thirst, salt and water balance, body temperature, gut movements and emptying of the bladder) or summarized it facilitates digestion, growth, immune responses and energy storage. The sympathetic nervous system (the other division) is the system for the four Fs, fight, flight, fright and sex. The sympathetic division mobilizes the body for a short-term emergency at the expense of processes that keep organism healthy over the long term [5, 14].

THE NEURONE DOCTRINE

The neurone doctrine asserts that nerve tissue is composed of individual cells called neurons that were genetic, anatomic, functional and trophic units. Wilhelm von Waldayer introduced the term "neuron" in 1891 and stated that nervous system consists of units that are not connected anatomically [25].

Fridtiof Nansen, neuroscientist, the arctic explorer and the humanitarian from Norway, who received the Nobel Prize award for peace in 1922, secured priority in establishing the neurone doctrine. His paper was published in 1886, where he reported that the cellular nerve units were not fused but only touched each other. He stated that the tangle of the nerve fibers is the true seat of the psycho. Nansen's concept of the cellular nerve unit as an independent entity provided a new view of the brain [26].

Interestingly, the founder of phychoanalysis, Sigmund Freud was the first one who, as a young medical student, pave the way for the neurone doctrine in 1877. He conceived the nerve cells and fibers are morphological and physiological units. Santiago Ramon y Cajal from Spain was rewarded for developing the neurone doctrine, which was established by Nansen two decades earlier. Cajal and Camillo Golgi from Italy were jointly awarded the Nobel Prize for physiology and medicine in 1906. In fact, progress in cellular neuroscience was not possible before the development of the compound microscope in the late seventeenth century. Observation of the brain tissue using microscope required thin slices in the range of diameter of the cells. Additionally, brain tissue has not firm consistency to make so thin slices. In the nineteenth century it was discovered how to harden ("fix") tissue by immersing it in formaldehyde. Microtome was developed to make very thin slices as well. The final breakthrough was the introduction of stains that could selectively color only some parts of the brain cells [5, 6, 26, 27, 28, 29].

Franz Nissl from Germany was introduced the stain in the late nineteenth century that distinguished neurons and glia from one another. Secondly, this stain enables histologists to study the brain cytoarchitecture or the arrangement of neurons in different part of the brain. Nissl stain showed nuclei of cells and the clumps of stained material around the cell nuclei called Nissl bodies. This stain presents neuronal cell body without its projections (neuritis). The Nissl-stained neuron looks like a lump of protoplasm containing a nucleus. In 1873, Camillo Golgi discovered that new silver stain which showed that neuron is much more than cell body. The neurons are showed to have at least two distinguishable parts: cell body (soma, perikaryon) with the nucleus and numerous thin tubes that radiate away from the body (neuritis: axons and dendrites) [5, 30].

Golgi established the reticular theory implying that the neurites of different neurons are fused together to form a continuous reticulum similar to blood vessels. According to Golgi, the brain is an exception to the cell theory which states that all animal tissues are built by cells as a elementary functional unit. Cajal, on the other hand, argued that the neurites of different neurons are not continuous with one another and they communicate by contact as a separate units. This idea, as we stated earlier, that the neuron adhered to the cell theory is known as the neuron doctrine [26, 30].

BRAIN ACTIVITY

Brain activity can be presented as several mental processes that all directing into the same goal. Brain functioning is in the form of distributed processing. Brain sciences and cognitive psychology show that all mental functions are divisible into subfunctions.

Canadian neurosurgeon Wilder Penfield created the homunculus. The most extensive studies of the effects of stimulating the cortex of the conscious patients was performed by Penfield (1930-1950). The studies showed that the different parts of the body were represented on the surfaces of the two gyri [31].

In the late 1950s, George Ojemann used small electrode to stimulate the cortex of awake patients during neurosurgical operations (under local anesthesia). They confirm in the living conscious brain the language areas of the cortex described by Broca and Wernicke. There is clear evidence for localized language-related functions in the brain cortex. Sensory, motor, cognitive functions are precisely mapped. Emotions was believed to be an expressed of whole-brain activity but emotions can be elicited by stimulating specific areas of the brain.

JAMES PAPEZ

The limbic system comprises structures interposed between the diencephalon (between-brain) and the cerebral hemispheres of the telencephalon (end-brain). Firstly, Paul Broca described the limbic lobe in

1878, and almost 60 years later James Papez (1937) stressed its role in emotional life. In fact, American neurologist Papez proposed that there is an "emotional system" on the medial side of the brain which links the neocortex (neopallium) with hypothalamus (Papez circuit) [14, 15, 21, 32].

Papez stated that the neocortex is critically involved in the experience of emotions. It is a fact that brain tumors or other structural lesions located in cingulate cortex are associated with emotional disturbances (fear, irritability, depression) [2, 32].

MacLean's Theory

Finally, the term "limbic system" was suggested by American physiologist Paul MacLean in 1952 as a designation for the limbic cortex and brain stem structures with which it has primary connections. MacLean's theory ("triune brain") proposed that the primate brain developed in three evolutionary phases: 1) *the reptilian brain* composed of the brain stem and striatal nuclei; 2) *the paleo-mammalian* brain that composed of rudimentary cerebral cortex of reptiles differentiated into the limbic lobe of lower mammals; 3) mass of cerebral cortex that increased to become the six-layered neocortex [2, 33].

This theory can be simplified and expressed as follows: a) Brain stem (reptile brain as a center of Freud's "Id" and Jung's collective unconscious); b) Limbic system and diencephalon (elementary emotions and motivations or "unconscious and conscious Ego"); c) Neocortex (conscious, logical thinking, inhibition, anticipation or "conscious Ego and Superego") [21, 34].

MacLean stated that the evolution of a limbic system enabled animals to experience and express emotions and emancipated them from the stereotypical behavior dictated by their brain stem.

Young (Adolescent's) Brain

Human brains undergo a notable reorganization between our 12th and 25th years. In fact, brain minimally grow in this period. Actually, the human brain reached 90 % of its size by the six years and a thickening skull is responsible for head growth afterward. Young people are still learning to use their brain's new established neuronal networks and they tend to make less use of brain regions than adults. Their inhibition is limited. But this "immature brain" is needed for decision to move from the safety of home into the complicated world outside.

But seeking sensation, which is characteristic of young human brain is not necessary impulsive. It can be planned and deliberate activity. Although sensation seeking can be dangerous, it can also produce new and positive things. Interest in novelty leads to useful experience and provides young people with intentions to penetrate into "terra incognita". Adolescents are the biggest risk-takers but they are not stupid. They use the same basic cognitive strategies that adults do and they are capable to recognize their mortal destiny. Young brains weigh risk versus reward differently than adult brains do. This is true especially if there is other brain around watching the risk-taker in dangerous situation. Succeeding often requires moving into new environment and less secure situations [5, 21, 35, 36].

Young brains, at a neural level, perceive social rejection as a threat to existence. Seeking sensation, novelty, risk and company of peers as the traits of adolescence brain make us more adaptive individuals and a species. Adolescence, as a key transitional period for the humans, forces us to leave a safe home and move into unfamiliar territory. Adolescence human brains are the most adaptive matters in our universe.

The speed of neural transmissions is directly connected with myelination but this myelin coating inhibits the growth of new branches from the axons and generations of new synapses. Different brain areas acquires its myelin insulation at different period of life, but once that is done, it is harder to be changed. Myelin insulation is done at the price of the flexibility of brain activities. It reduces plasticity of the brain which means reduction in the brain capability to acquire new knowledge. Brain's language areas acquire myelin insulation in the first 13 years during period of language learning but forebrain's myelination are completed during early 20s. This slow, back to front myelination, is uniquely human adaptation.

DIFFUSE MODULATORY SYSTEMS OF THE BRAIN

Neurons of the diffuse system mainly arise from the brain stem (reptile brain). Each neuron has an axon that can contact more than 100,000 postsynaptic neurons that have been distributed widely across the different parts of brain. Synapses within this system are able to release different neurotransmitters into the extracellular fluid so that they can diffuse to the many distant neurons.

The main modulatory systems use: norepinephrine (NE), serotonin (5-hydroxytryptamine, 5-HT), dopamine (DA) and acetylcholine (ACh) as a neurotransmitter [5, 6, 14, 21].

All of these transmitters activate specific metabotropic receptors (G-protein–coupled).

The paired locus coeruleus ("blue spot", from the Latin) is located in the rostral part of the pons (isthmus rhombencephali) and represent the norepinephrine center in the brain. It has some of the most diffuse connections in the whole brain considering that just one of its neurons can make more than 250,000 synapses. One axon of these neurons can have one branch in neocortex and another in cerebellar cortex.

Locus coeruleus is involved in regulation of attention, arousal, mood, pain, brain metabolism as well as learning and memory. But it does not run all show it is only involved in neural networks important for these processes. The main function of locus coeruleus is to increase brain responsiveness, especially for the unexpected, nonpainful sensory stimuli.

The nine serotonergic nuclei raphe (ridge, from the Greek) are located in the medulla, in the pons and midbrain. Each nucleus projects in the different areas of the brain and spinal cord in much the same diffuse way as do the locus coeruleus. Nuclei raphe are also important for arousal and they control the sleep-wake cycles.

The locus coeruleus (norepinephrine) and nuclei raphe (serotonin) are part of ascending reticular activating system (ARAS) which implicates the core of brain stem reticular formation.

ARAS arouses and awakens the forebrain. Serious damage of this system is always connected with deterioration of conscious state leading to coma. Neurosurgeon must protect patient's ARAS in the case of severe brain injury and the raised intracranial pressure accompanied by brain herniation.

The substantia nigra and ventral tegmental area are part of the midbrain and represent dopaminergic modulatory system. Neurons of substantia nigra project axons to the striatum (i.e., the caudate nucleus and the putamen) where they facilitate the initiation of voluntary movements. Parkinson's disease is result of neuronal degeneration in the substantia nigra. Neurons from ventral tegmental area innervate the frontal cortex and parts of limbic system. This projection is called the meso-cortico-limbic dopamine system and is involved in a "reward system" that reinforces certain adaptive behaviors. The meso-cortico-limbic dopamine system has been implicated in psychiatric disorders and cocaine addiction [5, 21].

Cholinergic modulatory system includes basal forebrain complex and ponto-mesencephalo-tegmental complex.

The basal forebrain consists of: the medial septal nuclei (innervations of the hippocampus) and the basal nucleus of Meynert (innervations of the neocortex). Neurons in the basal forebrain complex are among the first cells to degenerate during the course of Alzheimer's disease (profound loss of cognition). This complex is important in learning and memory.

The ponto-mesencephalo-tegmental complex provides a cholinergic link between brain stem and basal forebrain. It also projects to the dorsal thalamus where it regulates the excitability of the sensory relay nuclei.

DECADE OF THE BRAIN AND EVIDENCE BASED MEDICINE

The USA Congress designated the 1990s the "Decade of the brain". During the decade of the brain (1990-2000) effective therapy has been developed for multiple sclerosis, diabetic neuropathy, migraine and narcolepsy. A cochlear prosthesis has been perfected for congenital deafness. New antipsychotic, antiepileptic and antiviral drugs has been developed as well. The growth hormone replacement and growth hormone receptor antagonists have been started to use for acromegaly treatment.

Neurosurgeons have made significant changes in their daily work: endovascular therapy for the treatment of aneurysms and vascular occlusive disease, spinal implants and instrumentation, functional (molecular) imaging and imaging-guided surgery, deep brain stimulation for variety of movement disorders, advances in epilepsy surgery, concepts of minimal invasive neurosurgery [37].

With sophisticated computer-based imaging, neurosurgeons now have the ability to superimpose in three dimensional space the brain lesions.

Stroke prevention in the form of "brain attack concept" has helped heighten public awareness of stroke and treatment has become possible through appropriate and fast intervention.

Conceptual advances during the decade of the brain are very important. These include: a) stem cells in the nervous system; b) brain plasticity: c) mitochondrial disease; d) genetic basis of neurological disease and neoplasms; e) imaging of the central nervous system; f) neuroactive growth factors, peptides and cytokines.

Stem cells that were taken from subependymal zone of the adult brain underwent a temporal lobectomy for epilepsy demonstrates the ability to be cultured and to make synaptic connections. Mitochondrial genome can transmit neurological disease and many researches have focused on understanding of this process.

In the beginning of the 21st century some failures persist: a) gene therapy; b) treatment of central nervous system injury; c) transplantation and regeneration; d) stroke therapy especially the brain protection from ischemia; e) prenatal (fetal) surgery; f) glioma surgery; g) mind-brain relationships (as a the most important humanistic aspect of the neuroscience) [(6,37,38,39].

During the decade of the brain neurosurgeons experienced the coronation of prospective randomized clinical trail (RCT) that is a core of evidence based medicine (EBM).

EBM is the conscientious and judicious use of current best evidence in making decisions about the care of individual patients. It integrates clinical expertise with the best external clinical evidence from systematic research. EBM is derived from five ideas: a) clinical decision should be based on best available clinical evidence; b) clinical problem should determine the type of evidence sought; c) identification of the best evidence should be assessed using epidemiological and bio-statistical criteria; d) conclusion should be implemented in practice, e) result of decision (outcome) should be objectively evaluated [4, 40].

EBM needs clear clinical question, effective literature search, critical appraisal of evidence and clinical implementation. Fundament of EBM is that the peer-reviewed literature is the primary source for new clinical evidence. Common sense and reasoning from basic science pathophysiological principles in absence of confirmatory clinical empirical evidence, non-systematic personal experience without objective quantification, opinion of "experts" are all failed to qualify as a "real evidence" within EBM.

EBM de-emphasizes intuition, unsystematic clinical experience and pathophysiological rationale as sufficient grounds for clinical decision making. EBM divides clinical studies into those that address therapy, harm, diagnosis, and prognosis. For therapeutic questions, EBM insists on the priority of data derived from RCT, mega-RCT (more than 1000 patients) or meta-analysis of multiple RCT.

Within the realm of neurosurgery, a variety of reasons exist that limit neurosurgeons to perform RCT: 1) condition of interest is often rare, 2) some benign tumors (meningioma, schwannoma, adenoma) need long follow up (10 or more years), 3) few patients are willing to participate in RCT in which one group has surgery whereas the other group is managed by a less invasive method (Endovascular or Gamma Knife).

It is obvious that neurosurgeon often have to decide on the ground of poor evidence. The first clinical trial in neurovascular surgery and the first attempt at a RCT in operative neurosurgery was performed by Wylie McKissock et al. (1958-1965). Generally problem of EBM is that leading EBM experts tend to be academic clinical epidemiologists who no longer see patients as clinicians and who are no longer responsible for the patient outcome.

Neurosurgeons are the only neuroscientists who work with the human living brain and the anatomical and physiological substrate of the human mind. So neurosurgeons should play a pivotal role in directing the advances in neuroscience. Our ignorance of brain is still so vast that the "decade of the brain" should be re-designated into the "century of the brain" implying that we need a whole new century to understand major brain functions. It is important to stress that more US citizens are hospitalized with neurological and mental disorders than with any other major disease group, including heart disease and cancer.

NEUROSURGERY

Neurosurgery means that the living human brain operates the living human brain. Neurosurgery is particularly prone to ethical and legal criticism. The indication for any operation depend on the balance of the mortality and morbidity of the disease, the mortality and morbidity of the operation, and the prospects of relief or cure of the disease. If the immediate mortality and morbidity of operation exceeds the immediate morbidity and mortality of the disease we can talk about high risk (dangerous) neurosurgery. Morbidity in neurosurgery may well mean severe lasting disability. Neurosurgeon may conclude that morbidity is of greater significance than mortality.

In fact, high risk Neurosurgery often means high risk Neurosurgeon and can be equated with the dangerous Neurosurgeon.

Generally, indication for neurosurgical operation can be established only when the risk of operation is less than the risk of a conservative management as determined by the natural history of the lesion.

The specialty of neurosurgery is only partially related to the topographic diagnosis of the lesion, so-called "neurological surgery". In fact, the value of the neurological examination for topographic diagnosis is very limited. Current neurosurgery is based on respecting interrelated physiology and patho-physiology, which still cannot be fully visualized and understood.

Before Cushing, there was cranio-spinal surgery as a part of general surgery; afterward, integral physiological neurosurgery as a new specialty in surgery. The very special contribution of Cushing is that he established the respect for physiology and pathophysiology of the CNS. Especially, he considered that the hemostasis is important in surgery but the homeostasis is even more important and highly significant [2].

Mahmud Gazi Yasargil establised microneurosurgery. Microneurosurgery is not the occasional use of the operating microscope during exploration of the human brain. Microneurosurgery encompasses a cohesive concept comprising noninvasive approaches along the natural pathways of the cisternal systems to reach the lesions of the brain and to completely eliminate them, the goal being to achieve a pure lesionectomy. Microneurosurgery implies controled cisternal disection through microworld of our brain/central nervous system [2, 41].

Microneurosurgery requires intensive, long-term training mastering the details of neuroanatomy, acquiring proficiency in using bipolar coagulation and the high-speed drill, learning the microtechniques of dissection and repair of extra- and intracranial vessels and nerves, as well as pia, arachnoidea, and dura, and practicing the art of surgical approaches and techniques in the cadaver and animal laboratory.

Microneurosurgery will further evolve in the coming century, even if the current operating microscope design is replaced by a 3-D high-resolution imaging system coupled with diverse computers to assist and augment our information and to enhance brain-hand dexterity. At the moment, micro-neurosurgery is superior to all other contending specialties in neurotherapy. It offers to patients a proven and more effective therapy compared with other modalities. It shields patients from nonproductive academic discussion.

Neurosurgeons influence brain neuronal networks through deep brain stimulation, psychosurgery, neuromodulation and neuronal transplantation. Manipulations of the brain structural lesions are the traditional part of neurosurgical activities. These include microsurgery of extrinsic and intrinsic brain tumors, vascular lesion, infective disease, as well as treatment of the traumatic brain injury, pain, congenital disorders etc.

In 1949, the Portuguese neurologist Egas Moniz was awarded with the Nobel prize in medicine for the development of the prefrontal leucotomy [42, 43]. Ironically, this procedure is completely abolished today but his other major contribution (cerebral angiography) became the starting point for developping the other neurosurgical discipline - neurovascular (cerebrovascular) surgery.

Based on work of Ramon y Cajal, Moniz claimed that psychological life depended on the free flow of impulses through the synapses. Certain psychiatric patients had their mental life circumscribed to a limited cycle of thoughts and therefore symptoms could be alleviated by surgically interrupting these fixed connections. He also refers to the experimental works by Bechterew (dogs) and Fulton (chimpanzee).

The true pioneer of psychosurgery was Gottlieb Bruckardt from Switzerland who reported in 1882. that he excised portions of the cerebral cortex (topectomy) to treat psychiatric disorders [44]. But his work was not well received by the scientific community. Comparing with Bruckardt, Moniz was far less invasive and he did not refer on his work.

At the time Moniz introduced his technique, there was really little to be offered for the treatment of difficult psychiatric patients. In 1935, Moniz and neurosurgeon Almeida Lima designed the leucotome which consisted of a needle with a wire loop in the tip for cutting the white matter. Moniz reported the twenty operated cases the following year, seven of them were said to be cured and additional seven were improved. Real problem started with American involvement, when a psychiatrist Walter Jackson Freeman brought the leucotomy in USA starting to work with a neurosurgeon, James Watts.

Soon, Freeman disrupted this cooperation and proposed more extensive approach (called "frontal lobotomy"). In 1945, he adopted a trans-orbital approach (the ice pick technique for the frontal lobotomy) [45]. Until 1951, almost 20.000 operations were performed. Finally, the devastating results were achieved and this technique was abolished. At the end, Moniz was shot in the spine by a lobotomized patient [5]. But psychosurgery is still alive and present procedures are carried out based on careful evaluation of a multidisciplinary team (neurosurgeon, psychiatrist, clinical neurophysiologist).

Currently, there are a few neurosurgical procedures for psychiatric disorders (affective/anxiety/obsessive-compulsive disorders) as follows: anterior cingulotomy (anterior cingulum), subcaudate tractotomy (fronto-basal white matter), limbic leucotomy (anterior cingulum + fronto-basal white matter), capsulotomy (anterior limb of internal capsule).

Neuromodulation procedures are widely used in functional neurosurgery and epilepsy surgery [46]. Functional neurosurgery encompasses brain lesioning and deep brain stimulation (DBS) in order to alter pathological brain function. Deep stimulation of the brain combines stereotactic methods with implantation of electrical stimulators and an understanding of the functional anatomy of the deep brain nuclei. The stereotaxy has facilitated the development of two fields of neurosurgery: functional neurosurgery and stereotactic radiosurgery (Gamma knife) [47, 48]. Parkinson's disease as a important degenerative brain disease is surgically treated by DBS (subthalamic nucleus is used as a target) [49]. Chronic pain historically belongs to neuro-surgery (today in the realm of anesthesia). Percutaneous neuromodulation, peripheral nerve stimulation, motor cortex stimulation (MCS) are all included in neurosurgical neuromodulation.

Bilateral DBS of subthalamic nuclei for the surgical treatment of Parkinson's disease results in marked improvement in the motor function and quality of life. This treatment has become a popular with more than 40,000 implants worldwide. Stereotactic radiosurgery was conceived by Lars Leksell from Sweden in 1951. It allows the delivery of focal ablative lesions to the brain through a closed skull. In the future, DBS techniques could be replaced by molecular and cellular therapy for the Parkinson's disease.

Epilepsy surgery includes microsurgery (anterior temporal lobectomy and selective amygdalohippocampectomy) and neuromodulation (vagal nerve stimulation and more recently DBS). Surgery for temporal lobe epilepsy is not only safe but superior to prolonged medical treatment. But it has estimated that in the USA only 1,5% of eligible patients undergo surgery for epilepsy control.

High-frequency vagal nerve stimulation significantly reduces the seizure frequency including general and partial-onset seizures with alteration of awareness. Bilateral DBS of the anterior nuclei of the thalamus reduces seizure frequency as well.

Neural transplantation (molecular and cellular therapies) for the treatment of neurodegenerative disorders, stroke, neurotrauma is promising therapy [50]. Post-mortem study showed that fetal neural cells grafted into patients with Parkinson's disease can survive and function for over a decade in the host's brain. Transplanted fetal dopamine neurons can provide prolonged clinical benefits for the patients. The grafted neurons can develop occasional Lewy body-like inclusions over time without affecting graft function.

OPTIMISM

Optimism is the most important cognitive prejudice. Optimism bias is the belief that future will be much better than the past and present. This bias can lead us to miscalculations but also protects us and keeps us moving forward. Without optimism we might be cave dwellers. We are able to imagine better alternative realities and to believe that we can achieve them. Such faith motivates us to pursue our goals [22].

Optimistic people have connections with their intuition. When average human brains imagine the future even the most banal life events seem to take a dramatic turn for the better. It is obvious our tendency for optimism. The question is whether the human tendency for optimism is consequence of the structure and architecture of our brains?

Neural networks responsible for remembering episodes from our past might not have evolved for memory alone. The core function of the memory could be to imagine the future and prepare us for what has to come. Memory neural networks are not designed to perfectly reproduce past events but to help in constructing the future in our mind based on the past. It is a reason why some past details are deleted and others inserted in our mind when we try to recall past events.

The most extraordinary human brain capability is cognitive (mental) time travel or the ability to move back and forth in time and space in one's mind. We might think of events we experienced yesterday and then immediately of our plans for tomorrow. This capacity to envision a different time and place was crucial for our evolution and survival but it is also critical to our everyday life.

Cognitive time travel allows us to plan ahead. It is a result of prefrontal cortex's communicating with rostral anterior cingulate cortex, hippocampus and amygdala. But our conscious foresight has the big price – the understanding that in the future death awaits. The awareness of mortality on its own would have interfered with our daily functions. The only way for our brains to overcome this paradox was to emerge cognitive time travel together with irrational optimism. The envision of future relies partly on the hippo-campus which is important for episodic memory (memory for events, adventures and time-space relations). Damage of hippocampus is cause of two brain's deficits: memory deficit and unability to construct detailed images of future scenarios [14, 22, 32, 36].

NATURAL LAWS- CONCLUSION

We do not live in a chaotic universe because of the existence of natural laws. We live in deterministic system in which each segment will have some degree of uncertainties in its prediction.

The similar is with the brain. When we see our brain as a coherent whole system governed by its intrinsic laws rather than as a large number of disparate parts, the picture of our inner universe becomes more connected but not absolutely exact.

We create the chance for us in expanding universe only thanks to our brain.

REFERENCES

[1] 1. Perkel JM. *This Is Your Brain: Mapping the Connectome Life Science Technologies.* 18 January 2013. DOI:10.1126/ science.opms.p1300071.

[2] Yasargil MG. A legacy of microneurosurgery: Memoirs, lessons, and axioms. *Neurosurgery.* 1999; 45:1025-1091.

[3] Trefil J. *Cassell's laws of nature.* Wellington House. London. 2002.

[4] Bland M and Peacock J. *Statistical questions in evidence-based medicine.* Oxford University Press. 2002.

[5] Kandel E, Schwartz J, Jessell T. *Principles of neural science The McGraw-Hill Companies.* Fourth Edition. 2000.

[6] Bear MF, Connors, BW, Paradiso MA. *Neuroscience.* Second Edition. Lippincott Williams and Wilkins. 2001.

[7] Koso M, Dizdarevic K. Everyday Memory in Microsurgically Treated Patients After Subarachnoid Hemorrhage. *J Clin Med Res.* 2015;7:225-231.

[8] Koso M, Dizdarevic K. Attention and executive functions in micro-surgically treated patients after subarachnoid hemorrhage. *Asian j Neurosurg.* 2015;10:260-5.

[9] Bailey D, Geary D. Hominid brain evolution. *Hum Nat.* 2009;20:67-79. (doi:10.1007/s12110-008- 9054-0)

[10] Stephan H, Frahm H, Baron G. 1981 New and revised data on volumes of brain structures in insectivores and primates. *Folia Primatol.* 1981; 35:1-29. (doi:10.1159/000155963)

[11] Barton RA. 2007 Evolutionary specialization in mammalian cortical structure. *J Evol Biol.* 2007;20:1504-1511.

[12] Hirsch MC, Kramer T. *Neuroanatomy. 3D Stereoscopic Atlas of the Human Brain.* Springer. 1999.

[13] LeDoux J. *The Emotional Brain.* Weidenfeld and Nicolson. London. 1998.

[14] Glynn I. *An Anatomy of Thought.* Weidenfeld and Nicolson. London. 1999.

[15] Papez J.W. A proposed mechanism of emotion. *Arch NeurPsych.* 1937;38:725-743. doi: 10.1001/archneurpsyc. 1937.02260220069003.

[16] Loukas M, Pennell C, Groat C. Korbinian Brodmann (1868-1918) and his contributions to mapping the cerebral cortex. *Neurosurgery.* 2011;68:6-11; discussion 11. doi: 10.1227/NEU. 0b013e3181f c5cac.

[17] Arnold CL, Matthews LJ, Nunn CL. The 10k Trees website: a new online resource for primate phylogeny. *Evol Anthropol.* 2010;19:114-118. (doi:10. 1002/evan.20251)

[18] Allman J. Evolving Brains. *Scientific American Library.* New York. 1998.

[19] Gould E, Reeves AJ, Graziano MS, Gross CG. Neurogenesis in the Neocortex of Adult Primates. *Science.* 1999; 286:548-552.

[20] Pearce E, Stringer C, Dunbar RIM. New insights into differences in brain organization between Neanderthals and anatomically modern humans. *Proc R Soc B.* 2013;280:20130168.

[21] Greenfield S. *Brain Story.* BBC Worldwide Limited. 2000.

[22] Kahneman D. Thinking, *Fast and Slow.* 2011.

[23] Swanson LW. The Hypothalamus. In*: Handbook of Chemical Neuroanatomy.* Eds. Bjorklund A, Hokfelt T, Swanson LW. Elsevier. Amsterdam. 1987. pp. 1-124.

[24] Descartes, R. Treatise of Man. Translation by T.S. Steele. Harvard University Press, Cambridge. 1972.

[25] Winkelmann A. Wilhelm von Waldeyer-Hartz (1836-1921): an anatomist who left his mark. *Clin Anat.* 2007;20:231-4.

[26] Fodstad H, Kondziolka D, de Lotbiniere A. The Neuron Doctrine, the Mind, and the Arctic. *Neurosurgery.* 2000;47:1381-1389.

[27] Fine R. *A History of Psychoanalysis.* Columbia University Press. New York. 1979.

[28] Cajal, S. Ramon Y. Histology of the Nervous System of Man and Vertebrates. *Translation of Histologie du systeme nerveux de l'homme et des vertebres,* 1909-1911, by N. Swanson and L. W. Swanson. Oxford University Press. New York. 1955.

[29] Berciano J, Lafarga M, Berciano M. Santiago Ramón y Cajal. *Neurologia.* 2001;16:118-21.

[30] De Carlos JA, Borrell J. A historical reflection of the contributions of Cajal and Golgi to the foundations of neuroscience. *Brain Research Reviews.* 2007;55:8-16.

[31]   Kumar R, Yeragani VK. Penfield – A great explorer of psyche-soma-neuroscience. *Indian Journal of Psychiatry*. 2011;53:276-278. doi:10.4103/0019-5545.86826.

[32]   Goleman D. *Emotional intelligence*. A Bantam Book. 1995.

[33]   Newman JD, Harris JC. The Scientific Contributions of Paul D. MacLean (1913-2007). *The Journal of Nervous and Mental Disease*. 2009;197(1).

[34]   Freud S. *The interpretation of dreams*. Wordsworth Editions Limited. 1997.

[35]   Dobbs D. *Beautiful brains. National Geographic*. October 2011. pp. 36-59.

[36]   Morin E. *L'Homme et la mort*. Editions du Seuil. 1976.

[37]   Laws RL. The Decade of the Brain: 1990 to 2000. *Neurosurgery*. 2000;47:1257-1260.

[38]   Goldstein M. The Decade of the Brain: an era of promise for neurosurgery and a call to action. *J Neurosurg*. 1990;73:1-2.

[39]   Albert L, Rhoton Jr. AANS Presidential Address, 1990: Neurosurgery in the Decade of the Brain. *Journal of Neurosurgery. 1990*;73:487-495.

[40]   Pollock BE. An evidence-based medicine review of stereotactic radiosurgery. *Prog Neurol Surg*. 2006;19:152-170.

[41]   Dizdarevic K. *Aneurizmatska Subarachnoid hemorrhage: cerebral vasospasm and secondary cerebral ischemia*. Univerzitetski udžbenik. Gong. Sarajevo. 2011. pp.276.

[42]   Moniz E. Test surgical treatment of certain psychoses . *Bull Acad Med*. 1937;93:1379-1385.

[43]   Mashour GA, Walker EE, Martuza RL. Psychosurgery: past, present, and future. *Brain Res. Rev* 2005;48:409-419.

[44]   Burckhardt G. About Rindenexcisionen, as a contribution to the surgical treatment of psychosis. *Allg Z Psychiatr Med*. 1891;47:463-548.

[45]   Freeman W. Transorbital leucotomy. *Lancet*. 1948;2:371-373.

[46]   Engel J. Finally, a randomised controlled trial of epilepsy surgery. *N Engl J Med*. 2001;345:365-366.

[47]   Leksell L. Stereotactic radiosurgery. *J Neurol Neurosurg Psychiatry*. 1983;46:797-803.

[48]   Leksell L. The stereotactic method and radiosurgery of the brain. *Acta Chir Scand*. 1951;107:316-319.

[49] Rodriguez-Oroz MC, Obeso JA, Lang AE et al. Bilateral deep brain stimulation in Parkinson's disease: a multicentre study with 4 years follow-up. *Brain.* 2005;128:2240-2249.

[50] Dizdarević K, Dizdarević S. Transplantation in Central Nervous System. In: Karamehić J; Dizdarević Z (eds). *Clinical Immunology.* Svjetlost. Sarajevo. 2007. pp. 685-691.

In: Horizons in Neuroscience Research. Vol. 25   ISBN: 978-1-63485-286-9
Editors: A. Costa and E. Villalba   © 2016 Nova Science Publishers, Inc.

Chapter 2

THE ROLE OF THETA AND GAMMA OSCILLATIONS IN WORKING MEMORY CAPACITY

Anja Pahor and Norbert Jaušovec
University of Maribor, Maribor, Slovenia

ABSTRACT

Two experiments were conducted in order to investigate the relationship between theta and gamma frequencies, and the ratio between these frequencies, and working memory. In experiment 1, participants' resting EEG data was recorded after which they solved verbal and spatial tests of short term memory (STM) capacity. As predicted, theta/gamma cycle length ratio positively correlated with verbal STM capacity. In contrast, theta/gamma cycle length ratio positively correlated with the forward version of a spatial STM test and negatively correlated with the backward version of the test. Individual gamma frequency positively correlated with verbal and spatial STM capacities, whereas individual theta frequency only correlated with performance on the spatial STM tests. In order to verify the results of experiment 1, a second experiment was conducted in which either theta or gamma transcranial alternating current stimulation (tACS) was applied over parietal areas and contrasted against sham stimulation. Following sham or verum tACS, the participants solved a series of figural and verbal n-back tasks during which their EEG was recorded. The results showed that theta tACS significantly reduced the average reaction time on a difficult verbal n-back test, which was accompanied by changes in early visual processing. Collectively, these results provide further insight into the neural basis of working memory.

Keywords: theta frequency, gamma frequency, tACS, working memory, STM

1. INTRODUCTION

Human verbal short-term memory (STM) has a limited capacity and varies substantially between individuals. Miller (1956) famously proposed that it is limited to 7 ± 2 items. Short term memory is considered to be a component of working memory, which also contains non-storage components such as executive attention (Conway et al. 2002). Combined, these components enable a mental workspace in which information is temporarily stored and manipulated (Baddeley 2000; Logie et al. 2011). In the hippocampus, different spatial information is represented in different gamma band subcycles of a theta cycle (Lisman and Idiart 1995). Based on a review of neurophysiological studies focusing on memory processes, Lisman and Jensen (2013) proposed a theta-gamma neural code for the organization of multi-item messages. Lisman and Jensen's model has received support from rodent (Bragin et al. 1995; Fuchs et al. 2007; Senior et al. 2008; Colgin et al. 2009; Belluscio et al. 2012) and human studies (Axmacher et al. 2010; Holz et al. 2010; Schack et al. 2002). For example, Axmacher and colleagues (2010) showed that hippocampal theta – gamma cross-frequency coupling is associated with WM maintenance of multiple items and that the modulating theta frequency depended on memory load. Furthermore, Kamiński, Brzeziska, and Wróbel (2011) demonstrated that verbal short term memory capacity can be predicted by theta-to-gamma cycle length ratio; however, a significant relationship was found only at the Fz site. The authors also reported that individual theta and gamma cycle lengths predicted verbal STM. Further support for these findings comes from a recent modeling study of brain oscillatory activity and its relationship with WM capacity and reasoning abilities (Chuderski and Andrelczyk 2015). The model predicted individual differences in reasoning ability, the success of which depended on its theta-to-gamma cycle length ratio in the same way that the performance of participants depended on their WM capacity.

In order to determine whether the results of Kamiński et al. (2011) are replicable and can be extended to spatial STM, an experiment was conducted in which theta/gamma cycle length ratio and individual theta and gamma frequencies were correlated with performance on verbal and spatial tests of STM capacity. In the present study, a larger sample was used, which is more representative of the population and increases the likelihood of producing statistically significant results. We predicted that theta/gamma cycle length ratio would correlate both with verbal and spatial STM. Given that verbal and spatial STM share only 40% of their variance (Kane et al. 2004), we did not expect to find the same patterns of correlations for the two types of STM tasks. With the aim of further exploring the relationship between theta and gamma oscillations and working memory, a second experiment was conducted in which transcranial alternating current stimulation (tACS) was used to test whether stimulation in these frequencies affects performance on tests of working memory. tACS is a relatively new stimulation technique in which weak oscillating electrical currents are applied to the head (Thut and Miniussi 2009). tACS applied at conventional EEG frequencies entrains with or synchronizes neuronal networks (Antal and Paulus 2013) and it can be used to modulate brain oscillations in a frequency-specific manner (Herrmann et al. 2013). Evidence for this comes from an EEG study in which tACS applied at individual alpha frequency (IAF) increased individual alpha power, but did not affect lower and upper alpha power (Zaehle, Rach and Herrmann 2010). Using a combination of correlational and experimental research designs enabled us to explore the relationship between working memory, individual theta and gamma frequencies, and theta/gamma neural coding.

2. EXPERIMENT 1

In experiment 1 we sought to determine whether the finding that verbal STM capacity positively correlates with theta/gamma cycle length ratio (Kamiński et al. 2011) is replicable, and to extend these findings to spatial STM capacity. In addition, we examined the relationships between individually determined theta and gamma frequencies and STM capacity.

2.1. Procedure

Eighty students participated in the EEG study (40 males and 40 females; average age = 21 years and 3 months). Upon arrival at the lab, the participants were briefed about the nature of the study and were asked to complete an informed consent. None of the participants reported health problems or use of medication. Resting (eyes-closed) EEG was recorded for 3 minutes using a Quick-Cap with sintered electrodes. Using the 10-20 Electrode Placement System of the International Federation, EEG activity was monitored over 19 scalp locations (Fp1, Fp2, F3, F4, F7, F8, T3, T4, T5, T6, C3, C4, P3, P4, O1, O2, Fz, Cz, and Pz). All leads were referenced to linked mastoids and a ground electrode was applied to the forehead. Electrodes placed above and below the left eye were used to monitor vertical eye movements. The impedance was maintained below 5 kΩ. The digital EEG data acquisition and analysis system (SynAmps RT) had a bandpass of 0.15-100.0 Hz and a sampling rate of 1000 Hz.

2.2. Behavioral Testing

After the EEG recording session, the participants solved the forward and the backward versions of a Digit span test and a Corsi block tapping test. The order in which the two tests were presented was counterbalanced. The Digit span task measures verbal WM capacity (Wechsler 1981), whereas the Corsi block test provides a measure of spatial WM capacity (Corsi 1972). Both tasks were presented on a computer screen and were self-paced. Prior to performance on the task, the participants completed short practice sessions. In the forward Digit span task, the participants were presented with a stream of numbers which they had to immediately repeat (by typing on a keyboard) in the order of presentation. In the backward version of the task, the participants were asked to repeat the numbers in reverse order. In the Corsi block tapping test, the participants were presented with blue squares against a black background, some of which turned white in a particular sequence. Their task was to repeat the sequence by clicking on the squares in the same (forward version) or reverse order (backward version). As can be seen in Table 1, there were no significant gender differences in performance on the four tests. For this reason the whole sample was used for subsequent analyses.

2.3. Data Analysis

Individual theta and gamma frequencies, and the cycle length ratio between these frequencies, were determined for each EEG channel as proposed by Kamiński et al. (2011).

Table 1. Average performance and standard deviation on four working memory tests for males and females. DIGf = forward digit span test, DIGb = backward digit span test, CORf = forward corsi block test, CORb = backward corsi block test

	Males		Females		
Test	M	SD	M	SD	df (78)
DIGf	6.75	1.08	6.95	1.15	t= 0.80; p= 0.43; d=0.18
DIGb	6.58	1.71	6.93	1.47	t= 0.98; p= 0.33; d=0.22
CORf	6.06	0.96	5.93	0.96	t= 0.64; p= 0.52; d=0.14
CORb	5.66	1.09	5.65	0.96	t= 0.05; p= 0.96; d=0.01

The signal was filtered forward and backward in time (Hamming filter; filter order [FO]: 4096) in sequential bands for theta (4–5 Hz, 5–6 Hz, 8–9 Hz) and gamma (25–26 Hz, 26–27 Hz, 47–48 Hz) oscillations. In the next step, the envelope of each theta band was correlated (Pearson's r) with the envelope of each gamma band. The two frequency bands that had the highest positive correlation between the envelopes were defined as individual theta and gamma frequency bands in a given channel. The individual frequencies were used to calculate theta/gamma cycle length ratio for each channel (Kamiński, et al., 2011; Wróbel et al., 2007; Bekisz and Wróbel, 1999). EEG data was processed using EEGLAB toolbox (freely available from HYPERLINK http://www.sccn.ucsd.edu/ eeglab/) for MATLAB (The MathWorks, Natick, MA, USA).

2.4. Results and Discussion

Average theta/gamma cycle length ratios in 19 EEG channels were correlated with performance on working memory tests. Performance on the forward digit span significantly correlated with theta/gamma cycle length ratio at electrode F8 (r = 0.33, p = 0.003) while performance on the backward digit span correlated with theta/gamma cycle length ratios at electrodes F8 (r = 0.26, p = 0.02), C3 (r= 0.22, p = 0.047), and O1 (r = 0.27, p = 0.02). Figure 1 shows theta/gamma cycle length ratios at electrode F8 and forward digit spans of 80 individuals. Performance on the forward Corsi block test significantly correlated with theta/gamma cycle length ratio at electrode P4 (r = 0.27, p = 0.02) whereas performance on the backward Corsi block test showed significant correlations at electrodes F3 (r = -0.26, p = 0.02) and C3 (r = -0.28, p = 0.01). These results suggest that at certain electrode locations, theta/gamma cycle length ratio positively predicts performance on verbal STM tests and on the forward spatial STM test. However, the effect sizes are small to moderate (Cohen 1988). Similarly, Kamiński et al. (2011) reported that theta/gamma cycle length ratio positively correlated with performance on a forward digit span task at the Fz electrode. Our results further showed that theta/gamma cycle length ratio at electrodes F3 and C3 negatively correlated with performance on the backward spatial STM test.

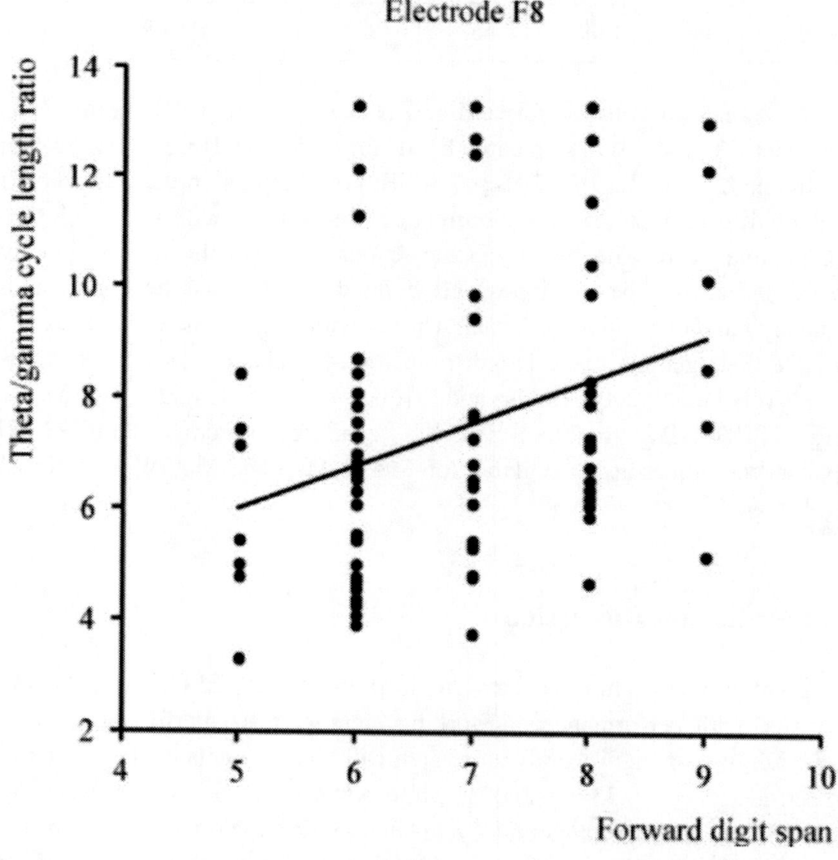

Figure 1. Scatter plot of verbal STM capacity (forward digit span test) and theta/gamma cycle length ratio at electrode F8 (N = 80).

The different correlations observed on the four STM tests support the view that verbal and spatial STM represent domain-specific storage mechanisms. We were also interested in examining the relationship between average individual theta and gamma frequencies in 19 EEG channels and performance on the STM tests. For theta, significant correlations were found only for forward (FP1: r = -0.25, p = 0.02) and backward Corsi block tests (F3: r = 0.26, p = 0.02; C3: r = 0.22, p = 0.05). In contrast, averaged gamma frequencies showed significant correlations with performance in three tests: forward corsi block (P4: r= 0.23, p= 0.04; F8: r= 0.30, p= 0.01); forward digit span (FP2: r = 0.26, p = 0.02; C3: r = 0.31, p = 0.01; O1: r = 0.24, p = 0.03; F8: r = 0.37, p = 0.001), and backward digit span (C3: r = 0.23, p = 0.04; O1: r= 0.40, p = 0.001; F8: r = 0.30, p = 0.01; Fz: r = -0.23, p = 0.04). In summary, theta correlated with performance on spatial STM tests in left frontal areas: it negatively correlated with the forward Corsi block test and positively correlated with the backward Corsi block test. In contrast, gamma correlated with performance on verbal and spatial STM tests, showing positive correlations across the whole head (with the exception of electrode Fz). Thus individual gamma frequency may serve as a stronger predictor of STM capacity than individual theta frequency.

3. EXPERIMENT 2

The results obtained in experiment 1 showed that individual gamma frequency is a stronger predictor of STM capacity than individual theta frequency. In order to verify this, a second experiment was conducted in which we investigated whether tACS applied in individually determined theta and gamma frequencies affected performance on a widely used test of working memory. Based on previous findings we predicted that verum stimulation would improve performance on the WM test (Santarnecchi et al. 2014; Jaušovec, Jaušovec and Pahor 2014; Pahor and Jaušovec 2014) and, with respect to experiment 1, that the most pronounced effect would be seen after gamma stimulation.

3.1. Procedure

Fifteen healthy female subjects participated in the second experiment (average age = 20 years and 10 months). Upon arrival they were briefed about the experiment and were asked to complete informed consents. The subjects participated in three experimental sessions which took place over three consecutive days. Each session involved (1) 3 minutes of resting eyes-closed EEG recording, (2) 15 minutes of tACS and (3) performance on working memory tests during which EEG was recorded. The EEG recording technique and the procedure used for the resting (eyes closed) condition were the same as in experiment 1 – for more information refer to section 2.1. Given that pre-stimulus theta oscillatory activity predicts successful memory retrieval (Guderian et al. 2009; Addante et al. 2011) and that pre-stimulus gamma oscillations predict the speed of reaction time (Gonzalez Andino et al. 2005; Womelsdorf et al. 2006), we decided to use the resting EEG data in order to determine individual tACS frequencies. These were obtained via the method described in section 2.3. Individual theta and gamma frequencies were determined only for the P3 and P4 channels. The frequency used for subsequent stimulation was the average of the individual frequencies determined in the two channels.

3.2. Electrical Stimulation

Transcranial alternating current stimulation was applied via two electrodes (7×5 cm) that were placed in saline-soaked sponges (DC-stimulator plus, Neuroconn, Ilmenau, Germany). The electrodes were positioned under an EEG cap at electrodes P3 and P4. The impedance was kept below 10 kΩ and the magnitude of the current was individually determined (Mode = 1750 μA; Range= 1500 - 2000 μA). Each individual participated in three experimental sessions: one sham and two verum, which were counterbalanced. tACS was administered for 15 minutes except in the sham condition in which stimulation ended automatically after 1 minute. For more information regarding the tACS procedure refer to previously published work (Pahor and Jaušovec, 2014). In the two verum conditions, tACS was delivered in individually determined theta and gamma frequencies (M_{theta} = 4.93; SD_{theta} = 0.90, M_{gamma} = 31.3; SD_{gamma} = 4.90).

3.3. Behavioral Testing

After stimulation the participants solved 6 versions of the n-back task while their EEG was recorded: figural and verbal variants of 1-back, 2-back, and 3-back tests. The order of the tests remained fixed, starting with a 1-back figural test and ending with a 3-back verbal test. The task items, which consisted of colored squares and two-letter syllables, were generated on STIM2 (Compumedics Neuroscan Systems, Charlotte, NC, USA) and appeared on the screen for 400 ms. The inter-stimulus interval was 2000 ms. The experiment began with a practice session in which the participants got acquainted with the task and with the response pad. A two-alternative forced choice design was used: the participants were presented with a stream of stimuli and were asked to press *1* on a response pad if the current stimulus matched the stimulus presented *N* items previously, or press *2* if the stimuli didn't match. The maximum correct score for each test was 50.

3.4. Data Analysis

Neuroscan software Version 4.5. (Compumedics, El Paso, TX, USA) was used to remove ocular artifacts from continuous (CNT) files. The digital EEG data acquisition and analysis system (SynAmps RT) had a band pass of 0.15–30.0 Hz (roll-off 24 dB per octave). The ERP analysis was performed on EEG data based on the common average reference. Epochs were extracted ranging from 200 ms before stimulus presentation to 1000 ms after its presentation. All epochs with amplitudes exceeding $\pm$ 100 µV were excluded from further analysis. The average voltages in the 200 ms pre-stimulus period served for baseline correction. Peak-to-baseline amplitudes and latencies were determined automatically using the following time windows: P1 (40–120 ms), N1 (120–220 ms), P300 (250–600 ms). In addition to analyzing phasic components, we examined tonic components in the ERP, which are often referred to as slow waves (Rösler, Heil and Röder 1997). Specifically, we determined the amplitude area in a late latency window, which ranged from 500 to 1000 ms (Azizian and Polich 2007).

3.5. Results and Discussion

3.5.1. Behavioral Data

Theta tACS

A GLM for repeated measures with the factors stimulation (sham/verum theta), n-back (1/2/3) and type (figural/verbal) was conducted on the number of correct answers on the n-back tests. There was a significant interaction effect between stimulation, n-back, and type ($F(2,28) = 4.35$; $p = 0.02$; $\eta^2 = 0.24$). Figure 2 shows that theta tACS tended to improve performance on verbal n-back tests and more generally on difficult n-back tests, probably due to ceiling effects observed on 1-back tests. However, no significant differences emerged for first order simple interaction effects between the factors stimulation and type for the three n-back tests (1-back: $F(1,14) = 0.07$; $p = 0.80$; $\eta^2 = 0.01$; 2-back: $F(1,14) = 2.92$; $p = 0.11$; $\eta^2 = 0.17$; 3-back: $F(1,14) = 1.52$; $p = 0.24$; $\eta^2 = 0.10$), thus we did not calculate second order simple effects. The effects of theta tACS were also examined with respect to reaction time.

The GLM showed a significant interaction between the factors stimulation and type ($F(1,14) = 9.23$; $p = 0.01$; $\eta^2 = 0.40$). Theta tACS significantly reduced reaction times in the n-back tests compared to sham, particularly in the verbal tests. As can be seen in Figure 2, paired samples t-tests showed that theta tACS (compared to sham stimulation) significantly decreased participants' reaction time in the verbal 3-back test ($t(14) = 2.93$; $p= 0.01$; $d = 1.57$).

Gamma tACS

A GLM with the factors stimulation (sham/verum gamma), n-back (1/2/3) and type (figural/verbal) was conducted in order to examine the effect of gamma tACS on performance on n-back tests. A significant interaction effect emerged between the factors stimulation and type ($F(1,14) = 4.51$; $p= 0.05$; $\eta^2 = 0.24$). Gamma tACS tended to increase the number of correct answers for figural stimuli and decrease it for verbal stimuli, as compared to sham stimulation.

First order simple interaction effects between stimulation and type were not statistically significant for 1-back ($F(1,14) = 0.70$; $p = 0.42$; $\eta^2 = 0.05$) and 2-back ($F(1,14) = 0.002$; $p = 0.97$; $\eta^2 = 0.001$) tests. A trend towards significance emerged for the simple interaction effect between stimulation and type for the 3-back test ($F(1,14) = 3.84$; $p= 0.07$; $\eta^2 = 0.22$), suggesting that gamma tACS only had a positive effect on the figural version of the 3-back test.

A GLM on the effects of gamma tACS on reaction time showed no significant main effects or interactions.

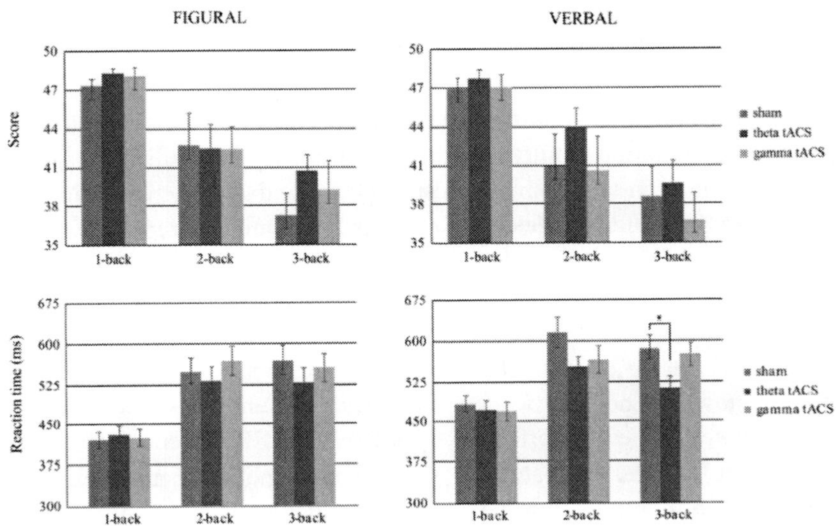

Figure 2. Average performance (number of correct answers) and reaction times for figural and verbal n-back tests in three sessions: sham, theta, and gamma tACS.

3.5.2. Physiological Data

Theta tACS

The effect of theta tACS on ERP amplitude was examined using a GLM for repeated measures with the following variables: stimulation (sham/theta verum), n-back (1/2/3), type (figural/verbal), and location (Cz/Fz/Pz). For P1 amplitude, the main effect of stimulation showed a trend towards significance $(F(1,14) = 4.01; p = 0.06; \eta^2 = 0.22)$ and a significant interaction effect between stimulation and type $(F(1,14) = 4.58; p = 0.05; \eta^2 = 0.25)$. Theta tACS increased the P1 amplitude during performance on the WM tasks; however, the increase was the more prominent for verbal compared to figural stimuli, which can be seen in Figure 3. There is a growing amount of evidence that working memory and selective attention represent overlapping cognitive domains (for a review see Gazzaley and Nobre 2012). Thus it is possible that the P1 amplitude modulation elicited by theta tACS reflected selective direction of attention towards the stimuli, which translated as improved performance on verbal WM tests. Similar findings have been reported by studies that investigate the effects of selective attention on early markers of visual processing during performance on memory tests. For example, Rutman et al. (2010) reported that during selective stimulus encoding of complex real-world objects, the P1 component correlated with the ability to successfully recognize relevant WM stimuli. They further suggested that early visual activity modulation yields more salient stimulus representations, which results in superior recognition ability. Moreover, Giuliano and colleagues (2014) conducted a study in which they investigated individual differences in visual WM capacity and auditory selective attention. The authors reported an early attentional modulation of auditory P1 in response to linguistic probes. Importantly, the attention effect correlated with performance on a test of visual WM capacity.

GLMs were also conducted on the effects of theta tACS on N1 and P3 amplitude, however, no significant main effects or interactions emerged. In the next step we examined the effect of theta tACS on ERP latency. The GLMs conducted on P1 and N1 latencies showed no significant main effects or interactions. In contrast, the GLM on P3 latency showed a significant interaction effect between the factors stimulation, n-back, and type ($F(2,28) = 3.27$; $p = 0.05$; $\eta^2 = 0.19$). For figural stimuli, theta tACS increased the P3 latency in the 1-back task but decreased the latency in 2- and 3-back tasks. For verbal stimuli, theta tACS decreased the P3 latency in 1- and 2-back tasks, yet increased it in the 3-back task. P3 latency is generally considered as an indicator of stimulus evaluation timing (Polich 2007). The present results lend support to the finding that P3 latency is sensitive to task processing demands (Daffner et al. 2010). In the aforementioned study, P3 latency increased as n-back difficulty increased.

In addition, P3 latency was shorter for high performers than for low performers. High performers could be likened to participants during theta tACS sessions in which shorter P3 latencies in 1- and 2-back verbal tests were associated with improved performance on these tests. The results for P3 latency changes in the verbal 3-back are less clear.

Gamma tACS

In order to examine the effect of gamma tACS on P1 amplitude, a GLM with the factors stimulation (sham/verum gamma), n-back (1/2/3), type (figural/verbal), and location (Cz/Fz/Pz) was conducted. Significant interaction effects emerged between the factors stimulation and type ($F(1,14) = 5.46$; $p = 0.04$; $\eta^2 = 0.28$) and the factors stimulation, type, and n-back ($F(1,14) = 4.96$; $p = 0.01$; $\eta^2 = 0.26$). For figural stimuli, gamma tACS increased P1 amplitudes only in the 2- and 3-back tests.

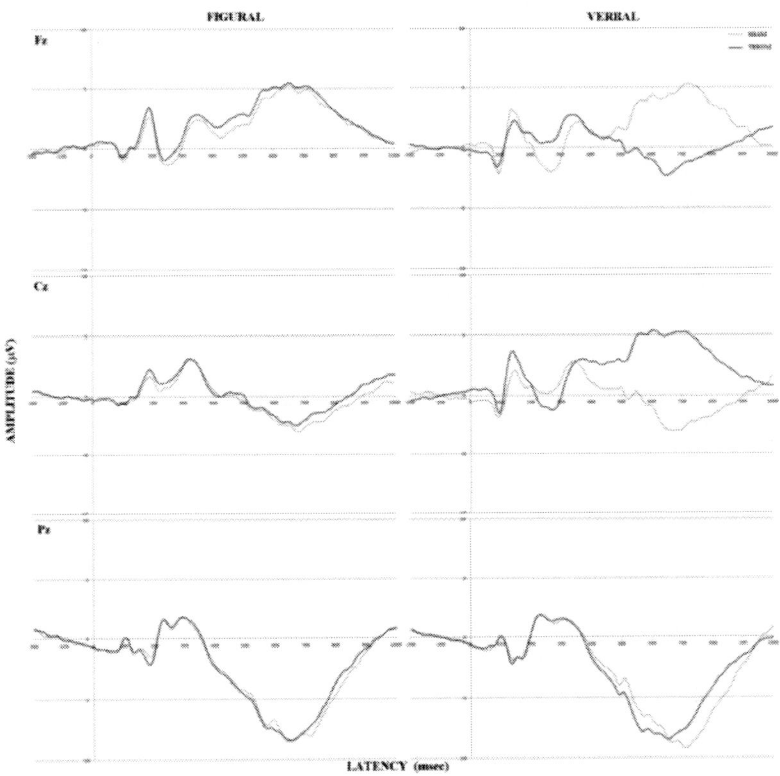

Figure 3. Grand averages for performance on figural and verbal n-back tests at electrodes Fz, Cz and Pz following sham and theta tACS.

These results are in line with the finding that gamma tACS somewhat improved performance on the figural 3-back test. For verbal stimuli, gamma tACS increased the P1 amplitude in all three tasks, with the strongest effect seen for the 1-back test; however, these changes did not translate into improved performance on the tests.

A GLM on the effect of gamma tACS on N1 amplitude showed no significant main effects or interaction. A GLM on P3 amplitude showed a trend toward significance for the interaction between stimulation, type, and location ($F(2,28) = 3.17$; $p = 0.057$; $\eta^2 = 0.19$). As can be seen in Figure 4, for figural stimuli, gamma tACS increased the P3 amplitude at Fz and Cz channels. For verbal stimuli, there was a large increase in P3 amplitude at the Fz channel and a decrease in P3 amplitude at the Pz channel. Subsequent analyses showed that gamma tACS had no effect on the latencies of P1, N1, and P3 components.

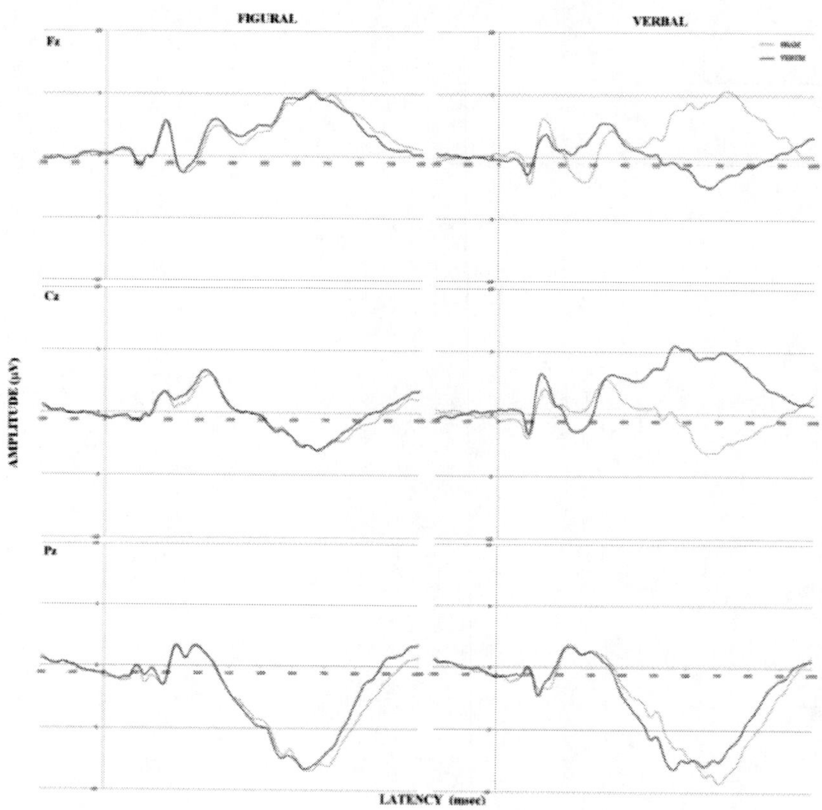

Figure 4. Grand averages for performance on figural and verbal n-back tests at electrodes Fz, Cz and Pz following sham and gamma tACS.

Late Latency Window (500-1000 ms)

The GLM for sham/verum theta tACS did not show any significant results. On the other hand, for sham/verum gamma tACS, the GLM showed a significant interaction effect between stimulation, n-back, type, and location ($F(4,56) = 6.42$; $p = 0.001$; $\eta^2 = 0.31$). Figure 5 shows the topographic distribution of the area voltage in the late window during sham and gamma tACS sessions. For figural stimuli, gamma tACS decreased the negative amplitude area over posterior areas, particularly in the 3-back test. For verbal stimuli, gamma tACS decreased the negative amplitude area over anterior areas– this effect was most prominent for 2- and 3-back tests. According to Rösler et al. (1997), slow waves reflect the allocation of "resources" to cortical processing modules.

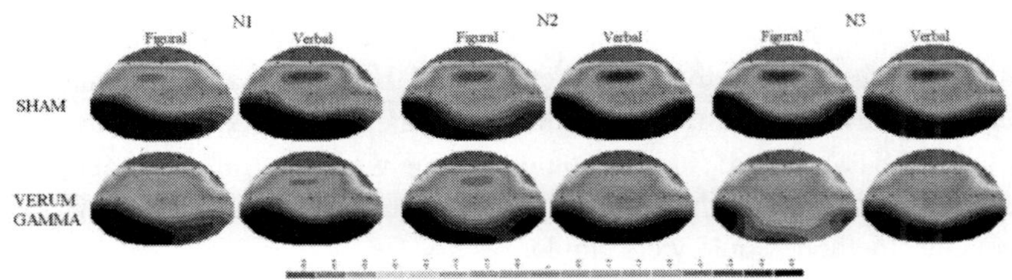

Figure 5. Topographic maps illustrating the area voltage distribution across all scalp electrodes from the late latency window (500–1000 ms) during performance on figural and verbal n-back tasks following sham and gamma tACS.

The topography of changes in slow waves shows where resources are allocated and the amplitude indicates the computational effort that is fed into a set of cortical cell assemblies. The authors also report that the amplitude of slow negative waves increases as task difficulty increases, and that these waves show task-specific topography. It is possible that the decrease in negative area voltage in gamma tACS sessions compared to sham sessions, which was observed in difficult (figural and verbal) n-back tests, reflects a reduction in computational effort.

CONCLUSION

In the first experiment, resting EEG data was used to determine individual theta/gamma cycle length ratios and individual theta and gamma frequencies. These values were then correlated with performance on tests of STM capacity. Individual theta/gamma cycle length ratio positively correlated with verbal STM capacity at electrodes F8, C3, and O1, which is line with the findings reported by Kamiński et al. (2011). In order to extend these findings, theta/gamma cycle length ratios were also correlated with performance on tests of spatial STM capacity. As expected, a different pattern of results was observed: the ratio positively correlated with the forward version of the task (electrode P4) and negatively correlated with the backward version of the task (electrodes F3 and C3). It should be noted that the obtained significant correlations had small to moderate effect sizes (refer to section 2.4.) thereby casting doubt on the reliability of this method to predict verbal and spatial STM capacities. In the next step, STM capacity was correlated with individual theta and gamma frequencies. While theta frequency only correlated with spatial STM capacity (in left frontal areas), gamma frequency positively correlated with verbal and spatial STM capacity across the whole head. These results suggest that individually determined theta and gamma frequencies, measured during a resting period, may be related to subsequent performance on tests of STM. Future studies should examine whether a different method for determining individual theta and gamma frequencies, such as the mean peak frequency method, would yield similar results.

Transcranial alternating current stimulation offers the possibility to demonstrate causal relationships between brain oscillations and cognitive processes (Herrmann et al. 2015). For this reason, a second experiment was conducted in which tACS was applied over parietal areas in individually determined theta and gamma frequencies. Following sham/verum stimulation, participants solved figural and verbal n-back tests while their EEG was recorded. Based on the results of experiment 1, it was predicted that gamma tACS would have the most pronounced effect on working memory performance. The results showed that theta tACS, as compared to sham stimulation, significantly reduced the average reaction time on the verbal 3-back test. On the other hand, gamma tACS tended to increase the number of correct answers on the figural 3-back test, with no significant effects on reaction time. The reason it was not possible to confirm our hypothesis that gamma tACS would have a stronger effect on WM performance than theta tACS might lie in methodological differences between the two experiments, which focused on STM and WM – of which WM is considered to be a more complex construct than STM (Conway et al. 2002). It would be interesting to conduct a similar tACS experiment with fixed frequency bands (5 Hz and 40 Hz, respectively) or with frequency bands determined based on individual alpha frequency. Nevertheless, the present results support previous findings on the positive effects of theta tACS on performance on memory tasks (Jaušovec, Jaušovec and Pahor 2014; Polania et al. 2012). Sauseng and collagues (2010) proposed that theta activity is crucial for the control of WM functions. The authors suggested that in WM, theta oscillations: act as a gating mechanism, define time windows optimal for encoding and retrieval of information, organize multi-item information into sequential WM representations, and connect neural structures by the means of interregional synchronization. It is likely that at least one of the above mechanisms can explain the improved performance on WM tests observed after theta tACS.

The ERP analyses showed that the changes in performance observed after theta and gamma tACS were mainly related to early visual processing. Namely, theta tACS increased the amplitude of P1 during performance on WM tests, particularly for verbal stimuli. While no significant differences were observed for theta tACS with respect to the amplitude of P3, which is thought to play a role in updating contents in WM (Polich, 2007), significantly shorter P3 latencies emerged for the 1- and 2-back verbal tests, indicating that the participants were able to evaluate stimuli in shorter periods of time.

Gamma tACS also resulted in an increase in the amplitude of P1 – the effect was strongest for figural 2- and 3-back tests. During verbal n-back tests, gamma tACS increased the P1 amplitude on the 1-back test, yet this was not associated with improved performance. Altogether, these findings indicate that the changes in performance on WM tests observed after theta and gamma tACS are at least in part a result of higher P1 amplitudes, which may reflect either enhanced visual processing or selective attention. It has been shown that attentional modulation of early visual processing relates to subsequent WM performance (Zanto and Gazzaley 2009). Both visual processing and successfully directing attentional resources to the fast-paced n-back task are crucial for good memory performance; however these cannot be distinguished based on the present results.

In the next step we examined the amplitude of slow waves (500-1000 ms) in sham and verum sessions. The results on the late latency window were significant only for the sham/gamma tACS comparison. Gamma tACS decreased the negative amplitude area over posterior areas for figural stimuli and decreased it over anterior areas for verbal stimuli. These effects were most prominent for difficult WM tests, implying that gamma tACS reduced the computational load required to solve the tasks.

In conclusion, the results of the first part of the study support the finding that theta/gamma cycle length ratio, derived from frontal electrodes, predicts verbal STM (Kamiński, et al. 2011). The relationship between theta/gamma cycle length ratio and spatial STM appears to be more complex, which may reflect the different mental processes involved in the forward and backward versions of the task. In addition, the present findings support the idea that verbal and spatial STM are domain-specific. The results further showed that average individual theta and gamma frequencies, measured during eyes-closed rest, correlated with performance on STM tests, however, further research is needed to validate this finding. The results of the second part of our study showed that tACS applied in the theta frequency reduced the average reaction time on a difficult verbal test of WM, which was accompanied by changes in early-stage visual processing. For gamma tACS, a trend towards significance was observed for the number of correct responses on a difficult figural test of WM. Collectively, these findings (1) provide new insight into the neural bases of STM mechanisms and (2) support the role of theta oscillations in working memory.

REFERENCES

Addante, Richard J., Andrew J. Watrous, Andrew P. Yonelinas, Arne D. Ekstrom, and Charan Ranganath. "Prestimulus Theta Activity Predicts Correct Source Memory Retrieval." Proceedings of the National Academy of Sciences 108, no. 26 (June 28, 2011): 10702–7. doi:10.1073/pnas.1014528108.

Antal, Andrea, and Walter Paulus. "Transcranial Alternating Current Stimulation (tACS)." *Frontiers in Human Neuroscience*, 7 (June 28, 2013). doi:10.3389/fnhum.2013.00317.

Axmacher, Nikolai, Melanie M. Henseler, Ole Jensen, Ilona Weinreich, Christian E. Elger, and Juergen Fell. "Cross-Frequency Coupling Supports Multi-Item Working Memory in the Human Hippocampus." Proceedings of the National Academy of Sciences, January 26, 2010, 200911531. doi:10.1073/pnas.0911531107.

Azizian, Allen, and John Polich. "Evidence for Attentional Gradient in the Serial Position Memory Curve from Event-Related Potentials." *Journal of Cognitive Neuroscience,* 19, no. 12 (November 13, 2007): 2071–81. doi:10.1162/jocn.2007.19.12.2071.

Baddeley, Alan. "The Episodic Buffer: A New Component of Working Memory?" *Trends in Cognitive Sciences*, 4, no. 11 (November 1, 2000): 417–23. doi:10.1016/S1364-6613(00)01538-2.

Bekisz, M., and A. Wróbel. "Coupling of Beta and Gamma Activity in Corticothalamic System of Cats Attending to Visual Stimuli." *Neuroreport,* 10, no. 17 (November 26, 1999): 3589–94.

Belluscio, Mariano A., Kenji Mizuseki, Robert Schmidt, Richard Kempter, and György Buzsáki. "Cross-Frequency Phase–Phase Coupling between Theta and Gamma Oscillations in the Hippocampus." *The Journal of Neuroscience,* 32, no. 2 (January 11, 2012): 423–35. doi:10.1523/JNEUROSCI.4122-11.2012.

Bragin, A., G. Jando, Z. Nadasdy, J. Hetke, K. Wise, and G. Buzsaki. "Gamma (40-100 Hz) Oscillation in the Hippocampus of the Behaving Rat." *The Journal of Neuroscience,* 15, no. 1 (January 1, 1995): 47–60.

Chuderski, Adam, and Krzysztof Andrelczyk. "From Neural Oscillations to Reasoning Ability: Simulating the Effect of the Theta-to-Gamma Cycle

Length Ratio on Individual Scores in a Figural Analogy Test." *Cognitive Psychology,* 76 (February 2015): 78–102. doi:10.1016/j.cogpsych.2015.01.001.

Cohen, Jacob. Statistical Power Analysis for the Behavioral Sciences. Hillsdale, N.J.: L. Erlbaum Associates, 1988.

Colgin, Laura Lee, Tobias Denninger, Marianne Fyhn, Torkel Hafting, Tora Bonnevie, Ole Jensen, May-Britt Moser, and Edvard I. Moser. "Frequency of Gamma Oscillations Routes Flow of Information in the Hippocampus." *Nature,* 462, no. 7271 (November 19, 2009): 353–57. doi:10.1038/nature08573.

Conway, Andrew R. A., Nelson Cowan, Michael F. Bunting, David J. Therriault, and Scott R. B. Minkoff. "A Latent Variable Analysis of Working Memory Capacity, Short-Term Memory Capacity, Processing Speed, and General Fluid Intelligence." Intelligence 30, no. 2 (March 2002): 163–83. doi:10.1016/S0160-2896(01)00096-4.

Daffner, Kirk R., Hyemi Chong, Xue Sun, Elise C. Tarbi, Jenna L. Riis, Scott M. McGinnis, and Phillip J. Holcomb. "Mechanisms Underlying Age- and Performance-Related Differences in Working Memory." *Journal of Cognitive Neuroscience,* 23, no. 6 (July 9, 2010): 1298–1314. doi:10.1162/jocn.2010.21540.

Fuchs, Elke C., Aleksandar R. Zivkovic, Mark O. Cunningham, Steven Middleton, Fiona E. N. LeBeau, David M. Bannerman, Andrei Rozov, et al. "Recruitment of Parvalbumin-Positive Interneurons Determines Hippocampal Function and Associated Behavior." *Neuron,* 53, no. 4 (February 15, 2007): 591–604. doi:10.1016/j.neuron.2007.01.031.

Gazzaley, Adam, and Anna C. Nobre. "Top-down Modulation: Bridging Selective Attention and Working Memory." *Trends in Cognitive Sciences,* 16, no. 2 (February 2012): 129–35. doi:10.1016/j.tics.2011.11.014.

Giuliano, Ryan J., Christina M. Karns, Helen J. Neville, and Steven A. Hillyard. "Early Auditory Evoked Potential Is Modulated by Selective Attention and Related to Individual Differences in Visual Working Memory Capacity." *Journal of Cognitive Neuroscience*, July 7, 2014, 1–9. doi:10.1162/jocn_a_00684.

Gonzalez Andino, Sara L., Cristoph M. Michel, Gregor Thut, Theodor Landis, and Rolando Grave de Peralta. "Prediction of Response Speed by Anticipatory High-Frequency (gamma Band) Oscillations in the Human Brain." *Human Brain Mapping,* 24, no. 1 (January 2005): 50–58. doi:10.1002/hbm.20056.

Guderian, Sebastian, Björn H. Schott, Alan Richardson-Klavehn, and Emrah Düzel. "Medial Temporal Theta State before an Event Predicts Episodic Encoding Success in Humans." Proceedings of the National Academy of Sciences 106, no. 13 (March 31, 2009): 5365–70. doi:10.1073/pnas.0900289106.

Herrmann, Christoph S., Stefan Rach, Toralf Neuling, and Daniel Struber. "Transcranial Alternating Current Stimulation: A Review of the Underlying Mechanisms and Modulation of Cognitive Processes." *Frontiers in Human Neuroscience*, 7 (June 14, 2013). doi:10.3389/fnhum.2013.00279.

Herrmann, Christoph S., Daniel Strüber, Randolph F. Helfrich, and Andreas K. Engel. "EEG Oscillations: From Correlation to Causality." *International Journal of Psychophysiology*, Accessed February 18, 2015. doi:10.1016/j.ijpsycho.2015.02.003.

Holz, Elisa Mira, Mark Glennon, Karen Prendergast, and Paul Sauseng. "Theta–gamma Phase Synchronization during Memory Matching in Visual Working Memory." *NeuroImage,* 52, no. 1 (August 1, 2010): 326–35. doi:10.1016/j.neuroimage.2010.04.003.

Jaušovec, Norbert, Ksenija Jaušovec, and Anja Pahor. "The Influence of Theta Transcranial Alternating Current Stimulation (tACS) on Working Memory Storage and Processing Functions." *Acta Psychologica*, 146 (February 2014): 1–6. doi:10.1016/j.actpsy.2013.11.011.

Kamiński, Jan, Aneta Brzezicka, and Andrzej Wróbel. "Short-Term Memory Capacity (7 ± 2) Predicted by Theta to Gamma Cycle Length Ratio." *Neurobiology of Learning and Memory*, 95, no. 1 (January 2011): 19–23. doi:10.1016/j.nlm.2010.10.001.

Kane, Michael J., David Z. Hambrick, Stephen W. Tuholski, Oliver Wilhelm, Tabitha W. Payne, and Randall W. Engle. "The Generality of Working Memory Capacity: A Latent-Variable Approach to Verbal and Visuospatial Memory Span and Reasoning." *Journal of Experimental Psychology: General 133,* no. 2 (2004): 189–217. doi:10.1037/0096-3445.133.2.189.

Lisman, J. E., and M. A. Idiart. "Storage of 7 +/- 2 Short-Term Memories in Oscillatory Subcycles." *Science,* 267, no. 5203 (March 10, 1995): 1512–15. doi:10.1126/science.7878473.

Lisman, John. "The Theta/gamma Discrete Phase Code Occuring during the Hippocampal Phase Precession May Be a More General Brain Coding Scheme." *Hippocampus,* 15, no. 7 (2005): 913–22. doi:10.1002/hipo.20121.

Lisman, John E., and Ole Jensen. "The Theta-Gamma Neural Code." *Neuron,* 77, no. 6 (March 20, 2013): 1002–16. doi:10.1016/j.neuron.2013.03.007.

Logie, Robert H. "The Functional Organization and Capacity Limits of Working Memory." *Current Directions in Psychological Science,* 20, no. 4 (August 1, 2011): 240–45. doi:10.1177/0963721411415340.

Miller, George A. "The Magical Number Seven, plus or Minus Two: Some Limits on Our Capacity for Processing Information." *Psychological Review,* 63, no. 2 (1956): 81–97. doi:10.1037/h0043158.

Pahor, Anja, and Norbert Jaušovec. "The Effects of Theta Transcranial Alternating Current Stimulation (tACS) on Fluid Intelligence." *International Journal of Psychophysiology,* 93, no. 3 (September 2014): 322–31. doi:10.1016/j.ijpsycho.2014.06.015.

Polich, John. "Updating P300: An Integrative Theory of P3a and P3b." *Clinical Neurophysiology,* 118, no. 10 (October 2007): 2128–48. doi:10.1016/j.clinph.2007.04.019.

Richardson, John T. E. "Measures of Short-Term Memory: A Historical Review." *Cortex,* 43, no. 5 (2007): 635–50. doi:10.1016/S0010-9452(08)70493-3.

Rösler, F., M. Heil, and B. Röder. "Slow Negative Brain Potentials as Reflections of Specific Modular Resources of Cognition." *Biological Psychology,* 45, no. 1–3 (March 21, 1997): 109–41.

Rutman, Aaron M., Wesley C. Clapp, James Z. Chadick, and Adam Gazzaley. "Early Top-down Control of Visual Processing Predicts Working Memory Performance." *Journal of Cognitive Neuroscience,* 22, no. 6 (June 2010): 1224–34. doi:10.1162/jocn.2009.21257.

Santarnecchi, Emiliano, Nicola Riccardo Polizzotto, Marco Godone, Fabio Giovannelli, Matteo Feurra, Laura Matzen, Alessandro Rossi, and Simone Rossi. "Frequency-Dependent Enhancement of Fluid Intelligence Induced by Transcranial Oscillatory Potentials." *Current Biology,* 23, no. 15 (August 5, 2013): 1449–53. doi:10.1016/j.cub.2013.06.022.

Sauseng, Paul, Birgit Griesmayr, Roman Freunberger, and Wolfgang Klimesch. "Control Mechanisms in Working Memory: A Possible Function of EEG Theta Oscillations." *Neuroscience and Biobehavioral Reviews,* 34, no. 7 (June 2010): 1015–22. doi:10.1016/j.neubiorev.2009.12.006.

Schack, B., N. Vath, H. Petsche, H. - G. Geissler, and E. Möller. "Phase-Coupling of Theta–gamma EEG Rhythms during Short-Term Memory Processing." *International Journal of Psychophysiology,* 44, no. 2 (May 2002): 143–63. doi:10.1016/S0167-8760(01)00199-4.

Senior, Timothy J., John R. Huxter, Kevin Allen, Joseph O'Neill, and Jozsef Csicsvari. "Gamma Oscillatory Firing Reveals Distinct Populations of Pyramidal Cells in the CA1 Region of the Hippocampus." *The Journal of Neuroscience: The Official Journal of the Society for Neuroscience*, 28, no. 9 (February 27, 2008): 2274–86. doi:10.1523/JNEUROSCI.4669-07.2008.

Thut, Gregor, and Carlo Miniussi. "New Insights into Rhythmic Brain Activity from TMS–EEG Studies." *Trends in Cognitive Sciences*, 13, no. 4 (April 2009): 182–89. doi:10.1016/j.tics.2009.01.004.

Womelsdorf, Thilo, Pascal Fries, Partha P. Mitra, and Robert Desimone. "Gamma-Band Synchronization in Visual Cortex Predicts Speed of Change Detection." *Nature,* 439, no. 7077 (February 9, 2006): 733–36. doi:10.1038/nature04258.

Wróbel, Andrzej, Anaida Ghazaryan, Marek Bekisz, Wojciech Bogdan, and Jan Kamiński. "Two Streams of Attention-Dependent B Activity in the Striate Recipient Zone of Cat's Lateral Posterior–Pulvinar Complex." *The Journal of Neuroscience,* 27, no. 9 (February 28, 2007): 2230–40. doi:10.1523/JNEUROSCI.4004-06.2007.

Zaehle, Tino, Stefan Rach, and Christoph S. Herrmann. "Transcranial Alternating Current Stimulation Enhances Individual Alpha Activity in Human EEG." *PLoS ONE*, 5, no. 11 (November 1, 2010): e13766. doi:10.1371/journal.pone.0013766.

Zanto, Theodore P., and Adam Gazzaley. "Neural Suppression of Irrelevant Information Underlies Optimal Working Memory Performance." *The Journal of Neuroscience,* 29, no. 10 (March 11, 2009): 3059–66. doi:10.1523/JNEUROSCI.4621-08.2009.

In: Horizons in Neuroscience Research. Vol. 25   ISBN: 978-1-63485-286-9
Editors: A. Costa and E. Villalba   © 2016 Nova Science Publishers, Inc.

Chapter 3

THE POTENTIAL USAGE OF PLASTIC CLIP IN CEREBROVASCULAR NEUROSURGERY

Samir Delibegovic[1] and Kemal Dizdarevic[2]
[1]Department of Surgery, University Clinical Center,
Faculty of Medicine, University of Tuzla,
Bosnia and Herzegovina
[2]Department of Neurosurgery, University Clinical Center,
Faculty of Medicine, University of Sarajevo,
Bosnia and Herzegovina

ABSTRACT

The fourth generation of clips used in neurosurgery are made of titanium and its alloys. Titanium clips are standard in cerebrovascular operative neurosurgery. Apart from this, plastic clips are not applied in neurosurgery at all, although they have some practical advantages. One well-known advantage is connected with their intrinsic diamagnetic and nonconductive characteristic, that results in fewer CT and MRI artifacts than titanium clips.

The small difference in the inflammatory reaction between the titanium clip and the plastic clip in the animal canine model demonstrates the possibility of the potential use of plastic clips in operative neurosurgery. Also a three-dimensional 'finite element model' for a Hem-o-lok clip, placed on the aneurysmal neck of an imaginary intracranial supraclinoid internal carotid artery, and analysis of the influence of blood pressure on the biomechanical characteristics of the Hem-o-lok clip showed that Hem-o-lok plastic meets the necessary criteria for intracranial application and aneurysmal clipping procedures. At a maximum blood pressure of 250 mm Hg, the clip did not open at all causing blood leakage in the perivascular space. Also, the output data did not show any tension above the clip material's breakdown limit.

Many characteristics of plastic clips, including weight, strength of closing, biocompatibility and low price, make them attractive and potentially suitable for usage in cerebrovascular neurosurgery. Typically, clips are repositioned several times during microsurgical clipping procedures, before their definite placement on the aneurysmal neck. Therefore, it is vital that they are easily placed and removed. Unfortunately, Hem-o-lok clips need another instrument for removal, which is a time-wasting procedure. It remains the task of the manufacturers to facilitate the application of these clips in the limited intracranial space. The potential usage of the plastic clips described in this study could open new clinical perspectives.

Keywords: titanium clip, plastic clip, neurocranium

INTRODUCTION

Approximately 2% of the general population suffers from an intracranial aneurysm, according to data based on autopsy and angiographic studies [1]. Although most intracranial aneurysms remain undetected, acute rupture, resulting in subarachnoid hemorrhage, can be a devastating consequence associated with 30%–67% mortality and 15%–30% morbidity [2, 3]. Recent advances in noninvasive imaging, including CT, MR imaging, CT angiography (CTA), and MR angiography (MRA), have increased clinicians' ability to diagnose patients with this condition.

Clinical management of unruptured aneurysm patients is a current topic of debate. Preventive treatment for unruptured aneurysms, such as neurosurgical clipping or endovascular coiling, may be recommended, depending upon the patient's family history, and the aneurysm's size, morphology, and location. Surgical clipping has the advantage of being a time-honored, durable and versatile method for treating most intracranial aneurysms [4].

SURGICAL CLIPPING

An aneurysm clip was first applied to a cerebral aneurysm in 1937 by Dandy [5]. He applied a flat, silver clip on an aneurysm of the internal carotid artery. After this, aneurysm surgery began to develop rapidly. Yasargil revolutionized vascular neurosurgery through numerous innovations of clips, improvement of instruments and the introduction of a microscope. Drake, Sugita and Sundt made further contributions to the development of clips [5].

The first clips were ferromagnetic, which even led to a fatal accident in an MR imaging system [6]. Therefore the introduction began of first generation clips made of non-ferromagnetic or weakly ferromagnetic materials. The second generation was made of a stainless steel alloy. The third generation clips were made of a cobalt-based alloy. The fourth generation clips used in neurosurgery are made of titanium and its alloys [7]. Today, titanium clips are standard in cerebrovascular operative neurosurgery.

Apart from this, plastic clips are not applied in neurosurgery (Figure 1) at all, although they have some practical advantages.

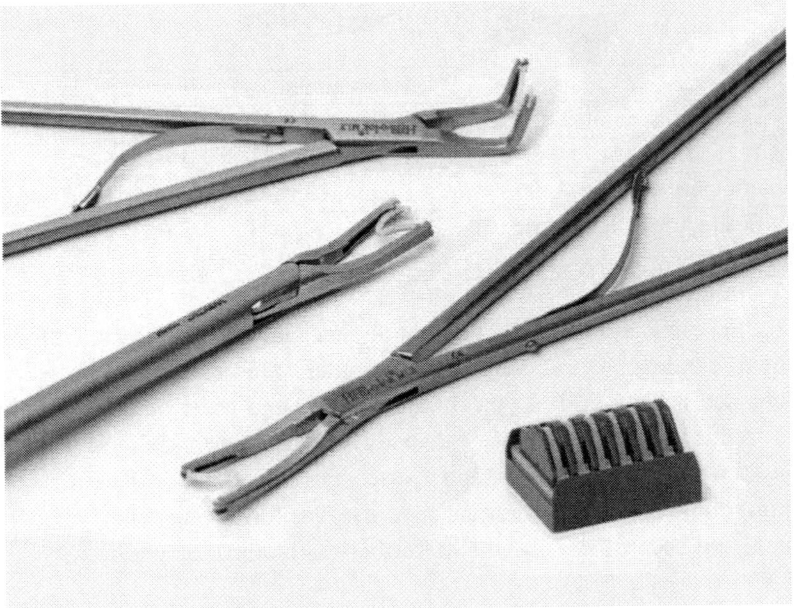

Figure 1. Medium-large Hem-o-lok clip and an applier.

Radiological Advantages of Plastic Clips

Two fundamental parameters characterize material behavior in X-ray imaging: effective atomic number and electron density (Table 1). The greater the electron density, the more interaction of X-ray photons with the sample material occurs.

These interactions may be absorption of the photons (removal from the beam) or scattering (change of direction with reduction in energy).

Homolymer polyacetal, from which Hem-o-lok clips are made, has a lower effective atomic number and density than a titanium clip and its alloys, which is why plastic clips cause fewer artifacts [8].

For postoperative evaluation of a clipped cerebral aneurysm, CT angiography is used, which is a time- and cost-saving investigation [9]. The potential use of plastic clips, which are cheaper and cause fewer CT and MR artifacts (Figures 2-5), enhances the time and cost saving aspects of their possible use.

Table 1. Comparison of the physical properties of titanium and polymer plastic clips

	Titanium clip	Polymer plastic clip
Effective atomic number	22	88
Density (g/cm3)	4.5	1.410-1.420
Magnetic permeability	1.25664133E-06	1,2566370639E-06
Specific electrical Resistance (μOhm-cm)	55	1.00

Plastic clips are made of homopolymer polyacetal, and therefore the electrical conductance is very low (normally the resistance to electrical conductance is very high. Typical values are in the order of 1.00e+13ohms), which reduces the occurrence of eddy currents. As such, it is a non-ferrous/non-magnetic material and causes fewer CT and MR artifacts than titanium clips. This is important because of follow–up after aneurysmal clipping, and because of subsequent, further radiological diagnostics.

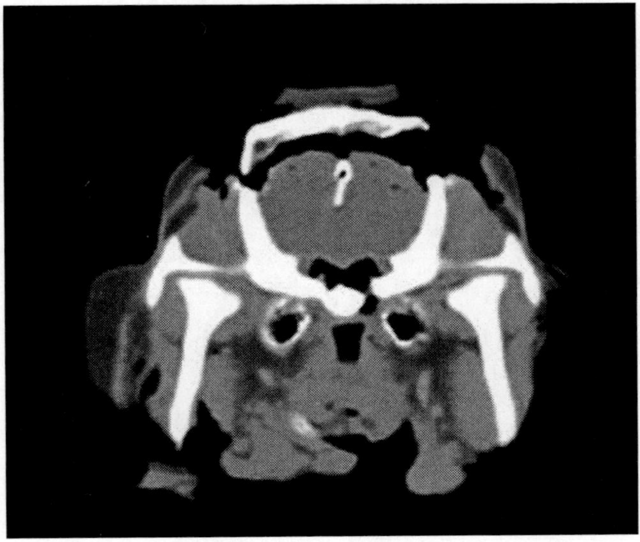

Figure 2. Computed axial tomography scan of a brain with a titanium clip (Sus scofa domestica).

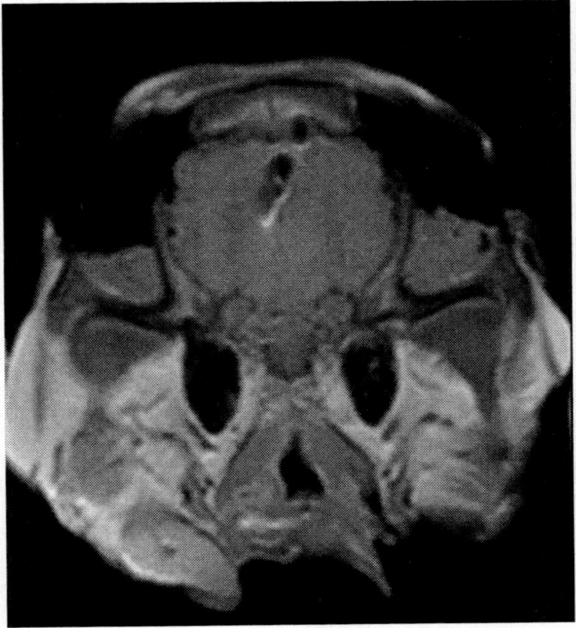

Figure 3. T1-weighted magnetic resonance imaging axial scan of the brain with a titanium clip (Sus scofa domestica).

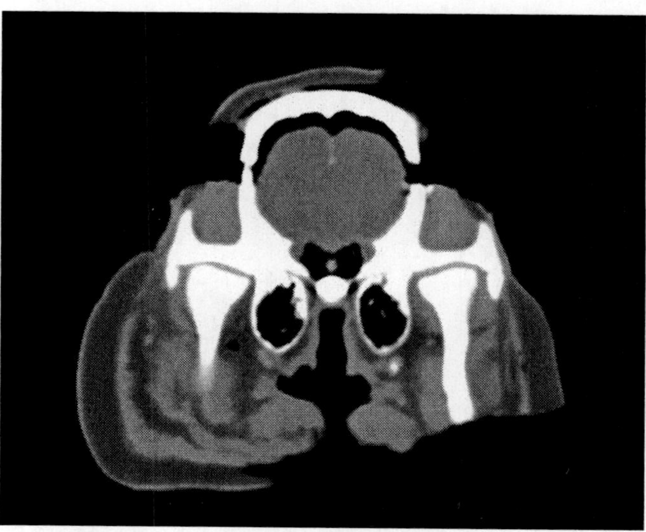

Figure 4. Computed axial tomography scan of brain with a Hem-o-lok ML plastic clip (Sus scofa domestica).

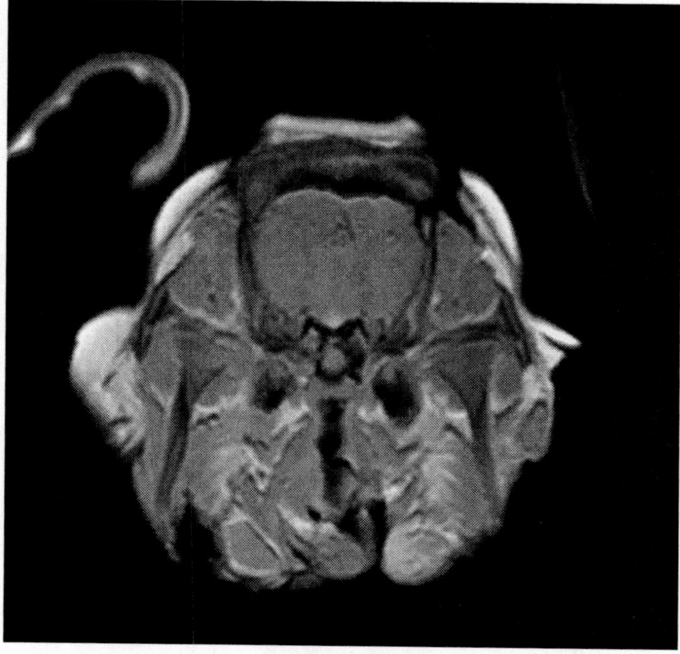

Figure 5. T1-weighted magnetic resonance imaging axial scan of the brain with a Hem-o-lok ML plastic clip (Sus scofa domestica).

Biocompatibility of Plastic Clips

Plastic clips have biocompatibility, which differs significantly from titanium clips. In our previous study [10], on the seventh postoperative day, in the groups with plastic clips, severe inflammation dominated, but there was no statistical significance in relation to the titanium clips. Generally, peak inflammatory reactions to foreign bodies occur on the seventh postoperative day [11]. However, even on the 60th postoperative day, there was still no significant difference in inflammation between the titanium and plastic clips.

The small difference in inflammation between titanium clips, which are the standard in neurosurgery, and plastic clips, demonstrates the possibility of the potential use of plastic clips in neurosurgery [10].

Endurance of Plastic Clips

A three-dimensional 'finite element model' for Hem-o-lok clip placed on the aneurysmal neck of an imaginary intracranial supraclinoid internal carotid artery and analysis of the influence of blood pressure on the biomechanical characteristics of the Hem-o-lok clip also showed that Hem-o-lok plastic meets the necessary criteria for intracranial application and aneurysmal clipping procedures. At the maximum blood pressure of 250 mm Hg, the clip did not open at all causing blood leakage in the perivascular space. Also, the output data did not show any tension above the clip material's breakdown limit.

Estimation of stress deformation condition was performed by nonlinear elasto-plastic analysis using ADINA R&D software. The finite element model consists of 5 groups of elements, and each group presents a detail of the clip system and vessel. The physical mechanical properties of the clip, that is modeled as an elastic bilinear element, are:

Modulus of elasticity: 1.6 N/mm^2, Poisson's coefficient: 0.4,
Yield stress σ_y =0.354 N/mm^2,
Modulus of elasticity: 0.00596 N/mm^2, Poisson's coefficient: 0.489

The model consisted of 23616 nodal points and 30272 elements and is presented in Figure 6. The behavior of the clip was simulated with pressure on the part of the vessel pinched with the clip, with increasing the pressure from 0 to double pressure than normal, in 100 steps (Figure 6).

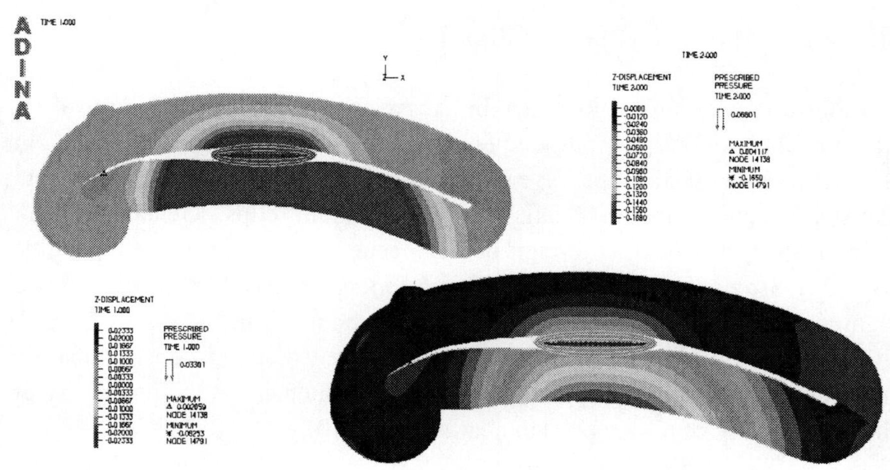

Figure 6. A three-dimensional 'finite element model' for Hem-o-lok clip.

Based on the results of the evaluation it may be concluded that the clip, due to its characteristics, wholly meets the criteria for its function. Namely, with a maximum blood pressure of 250 mm Hg, as with double pressure, there was no separation of the clip legs along the entire length of the clip, or creation of the free space for leaking of blood. Moreover, examination of the output data showed that there was no tension exceeding the clip material's breakdown σ_y.

CONCLUSION

The successful treatment of intracranial aneurysms requires a multidisciplinary approach, where surgery and endovascular therapies are viewed as complimentary instead of competing.

Imaging studies in patients with cerebral aneurysms are becoming increasingly sophisticated. This is why the radiological advantages of plastic clips may be of supreme importance for their future use in the neurocranium. Even in endo-vascular procedures, the possible use of a coil made of homopolymer polyacetal, may find a place.

Many characteristics of plastic clips, including weight, strength of closing, biocompatibility and low price make them attractive and potentially suitable for usage in cerebrovascular neurosurgery. Typically, clips are repositioned several times during a microsurgical clipping procedure before their definite placement on the aneurysmal neck. Therefore, it is necessary for them to be easily placed and removed. Unfortunately, Hem-o-lok clips need another instrument for removal, which is a time-wasting procedure. It remains the task of the manufacturers to facilitate the application of these clips in the limited intracranial space. The potential usage of the plastic clips described in this study could open new clinical perspectives.

REFERENCES

[1]   De Vries EN, Hollman MW, Smoenburg Sm, Gouma Dj, Boermeester MA. Development and validation of the SURgical PAtient Safety System (SURPASS) checklist. *Qual Saf Health Care*. 2009:18:121-126.

[2]   Hop J, Rinkel G, Algra A, et al. Case-fatality rates and functional outcome aftersubarachnoid hemorrhage: a systematic review. *Stroke*. 1997;28:660-64.

[3]   Raaymakers T, Rinkel G, Limburg M, et al. Mortality and morbidity of surgery for unruptured intracranial aneurysms: a meta-analysis. *Stroke*. 1998;29:1531-38.

[4]   Mason AM, Cawley III CM, Barrow DL. Surgical Management of Intracranial Aneurysms in the Endovascular Era: Review Article. *J Korean Neurosurg Soc*. 2009; 45:133-142.

[5]   Louw DF, Asfora WT, Sutherland GR. A brief historyof aneurysm clips. *Neurosurg focus*. 2001;11;E3.

[6]   Kluznik RP, Carrier Da, Pyka R, Haid RW. Placement of a ferromagnetic intarcerebral aneurysm clip in a magnetic field with fatal outcome. *Radioogy*. 1993;187:855-856.

[7]   Kakizawa Y, Seguchi T, Horiuchi T, Hongo K. Cerebral aneurysm clips in the 3-tesla magnetic field. Laboratory investigation. *J Neurosurg*. 2010;113:859-69.

[8]   Delibegovic S. Radiologic advantages of potential use of polymer plastic clips in neurosurgery. *World Neurosurgery*. 2014;3/4:549-551.

[9]   Zachenhofer I. Image quality and artifact generation post-cerebral aneurysm clipping using a 64 - row multislice computer tomography angiography (MSCTA) technology: a retrospective study and review of the literature. *Clin Neurol Neurosurg*. 2010;112:386-391.

[10] Delibegovic S, Dizdarevic K, Cickusic E, Katica M. Obhođaš M, Oćus
      M. Biocompatibility of plastic clip in neurocranium - Experimental
      study on dogs. *Turkish Neurosurgery*. 2016. In Press. DOI: 10.5137/
      1019-5149.JTN.13979-15.1.

[11] Baptista, ML, Bonsack ME, Felemovicius I, Delaney JP. Abdominal
      adhesions to prosthetic mesh evaluated by laparoscopy and electron
      microscopy. *J Am Coll Surg*. 2000;190:271-280.

In: Horizons in Neuroscience Research. Vol. 25   ISBN: 978-1-63485-286-9
Editors: A. Costa and E. Villalba   © 2016 Nova Science Publishers, Inc.

Chapter 4

SPINAL ROOT AVULSION: MOTONEURON DEATH AND STRATEGIES TO PROMOTE SURVIVAL OF INJURED MOTONEURONS

Heng Li[1], Carolin Ruven[1] and Wutian Wu[1,2,3,]

[1]School of Biomedical Sciences, Li Ka Shine Faculty of Medicine,
the University of Hong Kong, Pokfulam, Hong Kong SAR, China
[2]State Key Laboratory of Brain and Cognitive Sciences,
The University of Hong Kong, Hong Kong SAR, China
[3]Joint Laboratory for CNS Regeneration,
Jinan University and The University of Hong Kong,
GHM Institute of CNS Regeneration, Jinan University, Guangzhou, China

ABSTRACT

Spinal root avulsion is common in brachial plexus injury, with the rupture site located at the CNS/PNS transitional zone close to the motoneuron cell body. Progressive and rampant motoneuron death follows with a root avulsion, accompanied by profound changes of motonuerons in both cellular morphology and biochemical profile. Functional restoration firstly requires survival of the injured motoneurons. Different strategies have been reported to rescue the

---

* Coresponding author: Wutian Wu Email: wtwu@hku.hk.

lesioned motoneurons from death, which include: 1) Neurotrophic factors such as glial cell line-derived neurotrophic factor (GDNF) and brain-derived neurotrophic factor (BDNF) are efficient in promoting the survival of motoneurons. 2) Surgical repair, including implantation of peripheral nerve graft or reimplantation of the avulsed roots to the ventral spinal cord, could not only reduce motoneuron degeneration in both acute and delayed injury model, but also allow axons to regenerate into the nerve conduit. 3) With the coincidence of avulsion-induced cell death and the up-expression of nitric oxide synthase (NOS), manipulation of NOS is able to decrease motoneuron death. 4) As apoptosis has been indicated as one of the mechanisms leading to motoneuron death, inhibition of the caspase cascade promoted motoneuron survival. 5) In addition, successful axonal regeneration into the peripheral nerve trunk might be beneficial to the lesioned motoneurons. This chapter will review the avulsion-induced motoneuron death and discuss different strategies to facilitate survival of motoneurons following spinal root avulsion.

Keywords: spinal root avulsion, motoneuron death, neurotrophic factors, surgical repair, NOS

1. INTRODUCTION

Spinal root avulsion is the most serious, and unfortunately a common form of brachial plexus injury [1]. The lesion is located at the site where the spinal root is connected to the spinal cord. Therefore, this injury damages both the spinal cord and the spinal root [2].

In order to explain how the spinal root avulsion happens, it's necessary to depict its anatomical relationship between meninges and the spinal cord [1, 3]. The rootlets originate from the spinal cord and blend laterally to form the less numerous spinal roots. The ventral and dorsal roots extend through the intervertebral foramen before they coalesce and constitute the spinal nerve. The dura is the outer layer of meninges, which envelop the spinal cord. In the intervertebral foramen, it extends laterally to wrap the dorsal root ganglia and ventral root. As the spinal roots elongate more peripherally, the dura becomes the perineurium, while the epidural connective tissue is continuous with the epineurium. Therefore, the spinal roots are free to move in the intraspinal space. Upon a lateral traction, the spinal roots are the most vulnerable part because there is no protection from perineurium, which exists and provides sanctuary more laterally. Clinically, spinal root avulsion is found in violent traffic accidents, sports conflicts and difficult childbirths [2, 3]. Currently, there is still no cure to this type of injury and only palliative treatments are available [1].

Following an avulsion injury, there is rampant progressive motoneuron death [4], which can be attributed to several factors. Firstly, an avulsion injury deprives of neurotrophic factors provided by peripheral nerve and distal muscles [1]. Secondly, inflammatory processes involving microglia and cytokine activities also contribute to the death of injured motoneurons [5]. In addition, the possibly accompanied vascular trauma could lead to excitotoxicity. All those factors together result into loss of up to 90% of motoneurons [6].

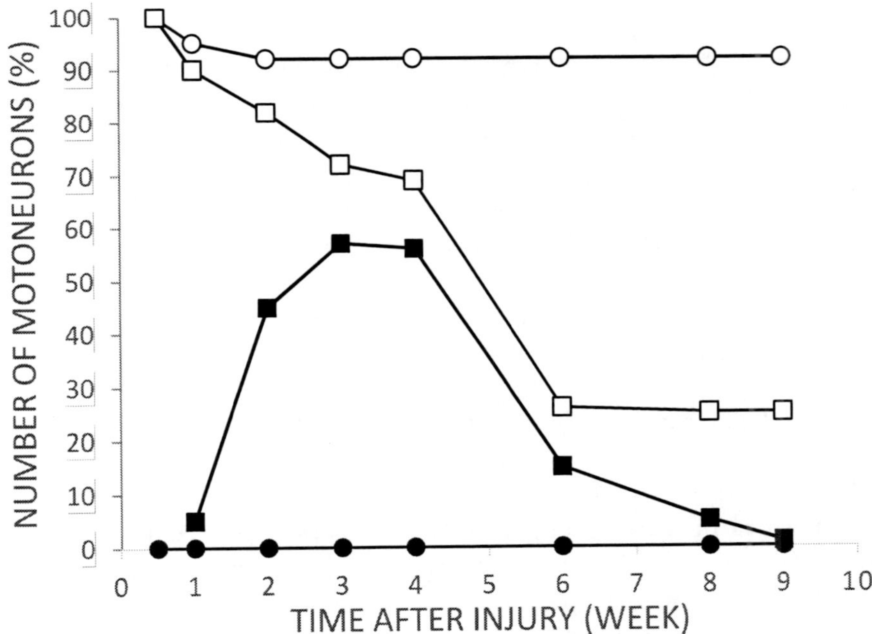

Figure 1. Total number of motoneurons and motoneurons expressing NADPH-d/NOS in the lesion side after the distal cut of C6 ventral roots and avulsion of C7 ventral roots. Values are expressed as percentages when compared with the total motoneurons in the intact side of the same spinal segment which is considered as 100%. O, total number of motoneurons in the lesioned side of the C6 spinal segment; □ total number of motoneurons in the lesioned side of the C7 spinal segment; ●, number of motoneurons expressing NADPH-d/NOS in the lesioned side of the C6 spinal segment; ■ number of motoneurons expressing NADPH-d/NOS in the lesioned side of the C7 spinal segment.

2. CHANGES OF MOTONEURONS FOLLOWING SPINAL ROOT AVULSION

2.1. Avulsion-Induced Progressive Motoneuron Death

Massive death of motoneurons is a distinctive feature of the spinal root avulsion, in contrast to the distal nerve axotomy. It has been shown that severe motoneuron death follows the complete spinal root avulsion [7], but it can be prevented when at least 4 mm of spinal root is left connected to the spinal cord [8]. This suggests that the remaining PNS stump helps to protect the motoneuron in the ventral spinal cord. The protection is most likely provided by the neurotrophic factors, produced by the peripheral nerve trunk and distal muscle fibers [9-13]. It's believed that motoneuron death after the avulsion is not immediate, but rather progressive and time-course specific. Namely, significant motoneuron death occurs from 2 weeks after the injury, and around 60-80% avulsed motoneurons stay alive at 2-4 weeks post injury. Dramatic decrease happens between 4 and 6 weeks, after which less than 30% of motoneurons can survive [7] (Figure 1).

2.2. Both Apoptosis and Necrosis Involved in Motoneuron Death

Although the mechanisms remain unclear, it's possible that both apoptosis and necrosis contribute to the motoneuron death after the avulsion injury [14-20]. For example, after the avulsion injury in adult rats, abnormal accumulation of metabolically active mitochondria in the perikaryon of the injured motoneurons has been observed [14, 21]. Chromatolysis has been detected and followed by progressive cytoplasmic and nuclear condensation, chromatin compaction into large round clumps, as well as rounding of the cells [14]. (Figure 2) Together with the positive TUNEL-staining and activation of caspases, these changes are typical to neuronal apoptosis [14, 15, 17]. On the other hand, some dying motoneurons exhibit non-apoptotic morphology, including dilated mitochondrial membrane, numerous vacuoles containing membranous debris in the cytoplasm and disintegrated cell membranes. In addition, the degeneration of oligodendrocytes and activation of microglia also suggests the involvement of necrotic pathway [15]. It's possible that necrosis has generally an early onset while apoptosis comes in at the later phase [14, 16, 18, 22].

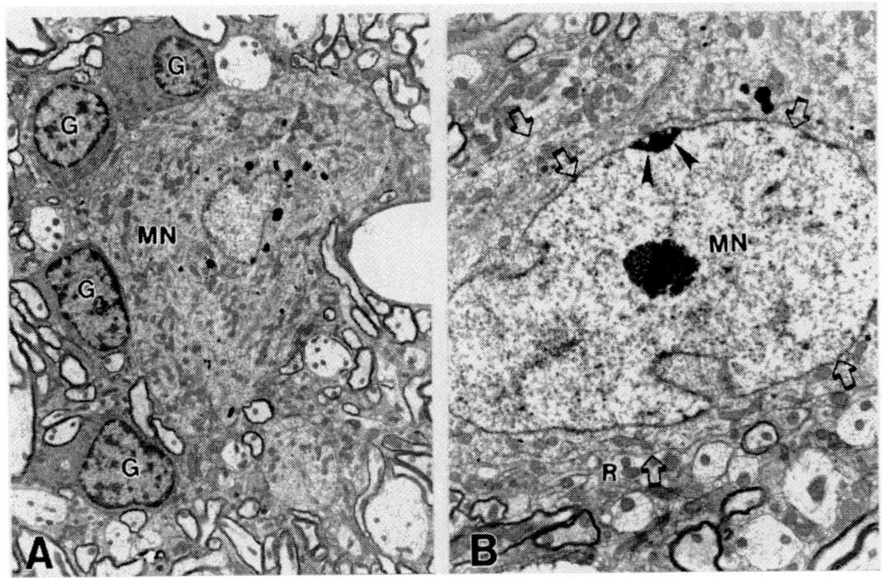

Figure 2. Electron micrographs showing morphological changes in rat spinal motoneurons 2 weeks after the root avulsion. (A): a shrunken motoneuron (MN) is surrounded by several glial cells (G). (B): the nucleus of an injured motoneuron (MN) is shrunken, but the cell and nuclear membranes are intact (arrows). An apoptotic body is found near the nuclear membrane (arrowheads).

2.3. Altered Gene Expression Profile Following Avulsion

The progressive nature of motoneuron death might be associated with the accumulative changes of expression profile after avulsion. Due to the complexity of expression profile, people have focused on some key genes that are involved in cell survival, apoptosis, regeneration etc. For example, as soon as 7 days after the avulsion injury, neuronal nitric oxide synthase (nNOS) is up-regulated in injured motoneurons, and the up-regulation stays for weeks [7] (Figure 1). The role of NOS in motoneuron death will be elucidated in section 3.3.1 of this chapter. Other genes related to apoptosis and DNA damage, such as ANX5, TS and ALR, are up-regulated after the avulsion injury [23, 24]. Meanwhile, genes responsible for cell survival, neuron protection and regeneration are generally down-regulated, such as IGFRII, PI3K, IGFBP-6, GSTs and GalR2 [23]. Nevertheless, lesioned motoneurons still fight for survival as they increase other set of protective genes, such as transcription factor c-Jun, neurotrophic factor receptor p75, c-Fos, GAP-43, and so on [25-28].

Our understanding on motoneuron responses to the avulsion is still very limited. In order to efficiently reverse the cell fate and eventually achieve survival, a lot more investigations are required.

3. Strategies to Promote the Survival of Motoneurons after Avulsion

3.1. Application of Neurotrophic Factors

3.1.1. Introduction to Neurotrophic Factors

Neurotrophic factors are classified as a group of molecules, which are essential for the survival of neurons, and their absence leads to the death of neurons. They are secreted to their targeted neurons and only tiny amount is available in living organisms [29, 30]. During development and adult life, neurotrophic factors play a critical role in the survival and maintenance of neurons. Since the first discovery of neurotrophic factor, i.e., the nerve growth factor (NGF)[31, 32], the list of this category has largely expanded and many members of this family have been identified, including brain derived neurotrophic factor (BDNF) [33], neurotrophin-3 (NT-3) [34, 35], glial cell line derived neurotrophic factor (GDNF) [36], ciliary neurotrophic factor (CNTF) [37] and so on.

Neurotrophic factors bind to two types of transmembrane receptors: the low-affinity receptor p75 and the receptor tyrosine kinases family (Trk) [38, 39]. Trk family members selectively bind to their ligands with high affinity, compared with the non-selective binding of p75. Neurotrophic factors exert their functions specifically to neurons or only to some subset of neurons. This results from the selective expression of receptors in different neurons and the binding specificity between ligands and receptors. For instance, while NGF was reported to promote the growth of sensory and sympathetic nerve cells [40], it has little effect on motoneurons [41], because the receptor of NGF, TrkA is rarely expressed by motoneurons. In order to search for neurotrophic factors that are protective to motoneurons, different types of them have been tested in the past decades and GDNF and BDNF have demonstrated high efficiency in rescuing motoneurons.

3.1.2. Protective Effects of GDNF and BDNF

GDNF was firstly purified from a rat glial tumor cell line, B47 [36]. *In vitro* experiment showed that GDNF treatment promoted the survival and morphological differentiation of dopaminergic neurons [36]. Since then, the effect of GDNF in promoting the survival of distinct types of neurons has been wildly investigated, either *in vitro* or *in vivo* [6, 42, 43]. GDNF demonstrated high efficiency in promoting motoneuron survival after spinal root avulsion injury [6, 24, 30]. Following a C7 spinal root avulsion, over 70% of injured motoneurons died at 6 weeks post surgery, whereas 92% of them stayed alive in rats with immediate GDNF treatment [24]. (Figure 3) Similar results were also observed at 4 weeks post surgery [30]. GDNF expression is not limited in the nervous system. Its mRNA was detected in the spinal cord, Schwann cells, as well as muscle fibers [44, 45]. In addition, following denervation, muscle fibers up-regulated GDNF level, indicating a possible role in axonal regeneration [45].

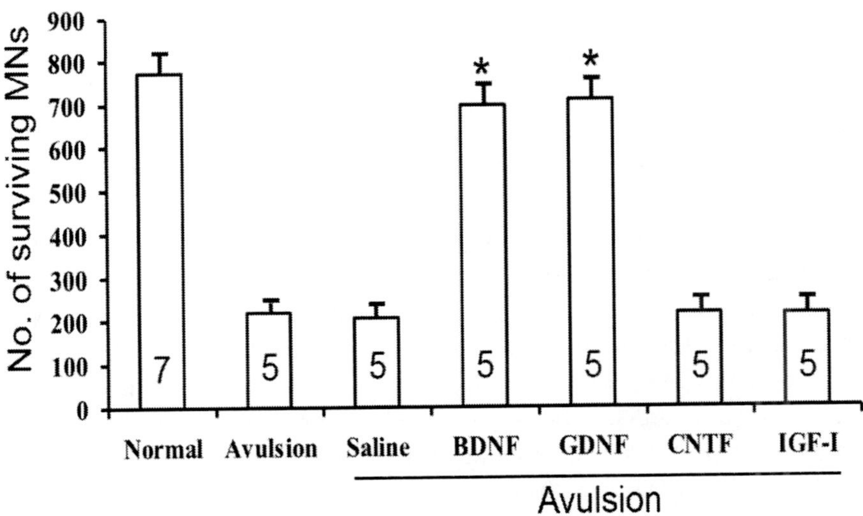

Figure 3. Neuroprotective effects of neurotrophic factors. All animals were subjected to C7 spinal root avulsion and 5 of the 7 groups (right side of the graph) were treated with saline, BDNF, GDNF, CNTF, IGF-1, respectively. The No. of surviving motoneurons (MNs) in the ipsilateral C7 spinal cord were counted at 6 weeks post surgery. Numbers in each bar represent the sample sizes of each group. *p < 0.001 when compared with saline treated group.

The first reported BDNF was purified from pig brain [46]. Not only showing the capability of supporting the survival and outgrowth of cultured sensory neurons from chick embryo [33], it also prevented the death of spinal motoneurons even at picomolar concentration [30]. Thereafter, the effect of BDNF in protecting motoneuron after injury was validated by different research teams [47-49]. Although not as effective as GDNF, BDNF still rescued 90% of injured motoneurons from death, at 6 weeks post avulsion [30] (Figure 3). BDNF binds to both p75 and TrkB. By activating TrkB, BDNF promoted survival and enhanced synaptic plasticity [39]. With BDNF expression detected in developing and adult spinal motoneurons [50], the expressions of both BDNF and TrkB were augmented within 7 days after the axonal injury [51]. In order to restore BDNF signaling, several types of TrkB agonists, with the capability of penetrating the blood-brain barrier, were developed [52-54]. These agonists demonstrated neuroprotection in neurodegenerative diseases and neurotrauma [52]. It's possible that promising TrkB agonists are also able to enhance motoneuron survival.

3.1.3. Dosage and Application of Neurotrophic Factors

Despite the facts that avulsed motoneurons die progressively and this process can be prolonged to as long as 6 weeks [24], immediate or earlier treatment of GDNF or BDNF is more beneficial than delayed administration. While invention with GDNF at 2 weeks after avulsion increased surviving motoneurons from 27.9% in control group to 86.1%, more delayed application at 4 weeks post injury failed to rescue the neurons [55]. In addition, single dosage of BDNF applied immediately was sufficient to display the long-term protection [24]. As neurotrophic factor is feathered by its ability of nourishing neurons at a low quantity, it was illustrated in rat that only 500 nanogram of GDNF was able to significantly prevent motoneuron death and several microgram of BDNF can achieve the comparable effect [30]. Another fact that makes the two neurotrophic factors promising is that the drug delivery can be simple. Gelfoam soaked with neurotrophic factor was put in contact with the injured spinal cord, in close proximity to the motoneuron pool [24, 30, 49]. In addition, it seems that both GDNF and BDNF can alter the expression of NOS and p75 [24] which are proteins involved in injury-induced neuron death [56]. However, the major challenge lies in that their capabilities of penetrating the blood-brain barrier are minimal.

3.2. Surgical Repair to Promote Motoneuron Survival

3.2.1. Implanting Peripheral Nerve Graft

Surgical approaches for root avulsion are limited when compared with those for distal nerve injuries, because the damage occurs on the interface of PNS and CNS [57, 58]. As an option, peripheral nerve graft can be used to connect the injured spinal cord and the peripheral nerves (PN). It has been shown that peripheral nerve grafts were able to enhance the motoneuron survival by being a source of neurotrophic factors [59-64]. Precisely, while only 30% of motoneurons survive at 6 weeks after the avulsion injury, PN graft promoted survival of more than 60% of avulsed motoneurons [60, 61, 72]. (Figure 4) Moreover, some of the injured motoneurons were able to regenerate their axons into the PN graft.

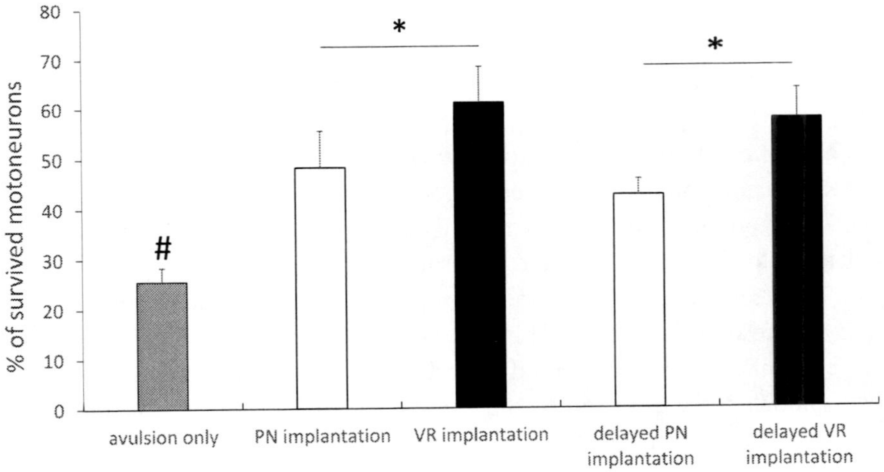

Figure 4. Both immediate and 1-week delayed peripheral nerve (PN) graft transplantation and ventral root (VR) re-implantation significantly increased the survival rate of motoneurons compared to control group at 6 weeks post avulsion. Furthermore, the survival rate of motoneurons in the animals receiving VR re-implantation was significantly higher than that seen in the animals receiving PN graft transplantation (#: p < 0.001 control group compared to PN or VR immediate or delayed implantation; *: p < 0.05).

As there is always a delay in clinic between the occurrence of the trauma and the surgery, some studies have compared the results of immediate and delayed implantation of nerve graft after the avulsion injury. After 20 weeks post avulsion, only 30% of injured motoneurons survived if no repair was applied, whereas 78% of them were alive in the immediate implantation group. When the implantation was delayed for 1, 2 or 3 weeks, around 65%, 57% or 53% of motoneurons survived, respectively, at the comparable time point. However, the regeneration capability of surviving motoneurons did not decrease with the delay [61]. Different animal studies have also shown some functional recovery after the peripheral nerve implantation [59, 61-63].

However, there are some drawbacks in using peripheral nerve graft for the repairment. Firstly, getting the peripheral nerve graft from the same patient will cause additional injury in another part of the body. Secondly, injured motoneurons have to regenerate their axons through two separate junctions: firstly from the CNS to peripheral nerve graft and then from the peripheral nerve graft to the distal nerve stump. Lastly, not all peripheral nerve trajectory can provide a suitable microenvironment for regeneration. As a matter of fact, implanting a pure motor nerve graft has shown to be better for motoneuron survival and regeneration, compared to the implantation of pure sensory nerve. This difference might be accounted for by the higher expression of GDNF and BDNF in the motor nerves than in the sensory nerves [65].

3.2.2. Reimplantation of Avulsed Ventral Roots

Reimplantation of avulsed ventral roots directly to the lesioned spinal cord is another potion. Mostly, this technique can be conducted by making an incision into the spinal cord and adhering the avulsed roots into the incision site. This technique has to be very carefully performed to avoid any further injury to the spinal cord [66-71]. Apart from the inevitable mechanical damage to the spinal cord, avulsed ventral roots also retract with time and might not be long enough for connecting to the spinal cord after a delay. Therefore, the approach has been optimized by placing the avulsed ventral roots ventrolaterally on the pia surface of the spinal cord, without making any incision [68, 72-79]. This can not only avoid damaging the spinal cord, but also overcome the problem of too short avulsed ventral roots.

In the adult rats with avulsion+reimplantation, around 90%, 80%, 62% and 55% of motoneurons survive at 3, 6, 12 and 20 weeks after injury, respectively, compared to the 65%, 39%, 14% and 11% of survival rate in rats without the reimplantation. (Figure 4) Also, around 80-90% of the surviving motoneurons have regenerated their axons into the reimplanted ventral roots [68, 71, 75]. These regenerated axons can form new neuromuscular junctions with target muscles and stimulation of these ventral roots can evoke the electromyography responses [70, 73, 75, 76]. With immediate and 2-weeks delayed reimplantation after avulsion, both animals and humans were able to regain some motor function [73, 75-77, 80, 81].

3.3. Others

3.3.1. Inhibition of NOS to Promote Motoneuron Survival

3.3.1.1. Nitric Oxide and Neuron Death

Nitric oxide (NO) is a free radical produced in some physiological conditions. The first report about this diffusible gas can be traced back to 1980's when people found that, originated from endothelium cells of blood vessels, it was able to mediate the relaxation of smooth muscle [82]. Later, publications about its roles as neuronal messenger [83-86] and the involvement in neuron degeneration [87] came out. It seems that NO is associated with both necrosis- [88, 89] and apoptosis- [90] induced neuron death. It's clear that high level of NO is toxic to neurons because as a free radical, NO together with its derivatives inhibits cytochrome oxidase, disrupts mitochondrial oxygen consumption and eventually damages the cellular generation of ATP [89].

3.3.1.2. NOS and the Avulsion-Induced Expression in Motoneurons

NO is produced by nitric oxide synthase (NOS) from the substrates of L-arginine and oxygen [91]. There are generally three NOS isoforms, neuronal NOS (nNOS), endothelial NOS (eNOS) and inducible NOS (iNOS). nNOS release basic level of NO in several types of neurons. However, increased NO production can be induced under pathological conditions, such as axonal injury, neural degenerative diseases, ischemic-hypoxic injury etc. [92] Because of the short life cycle and highly chemical reactivity of NO, most studies about its generation and localization have focused on NOS. Following a spinal root avulsion injury in adult rat, the expression of nNOS was induced in the injured motoneurons in the first week of injury [7, 56]. The amount of NOS increased rapidly at 2-4 weeks after injury (Figure 5) and up to 80% of motoneurons displayed NOS immunoreactivity at 4 weeks post avulsion [92]. Importantly, those motoneurons with NOS expression eventually die [92]. In addition, the avulsion-induced NOS expression in motoneurons varies with the age of the injured animals and could be distinctive among different species. Immature neurons are more vulnerable to axonal injuries, as shown in various types of injury models [92]. Consistently, young postnatal rats displayed faster and more intense NOS induction compared to the adult rats [7, 93]. Also, even though spinal root avulsion causes serious motoneuron death in both rats and mice, the *de novo* expression of NOS was only detected in rats but not in mice [92, 93].

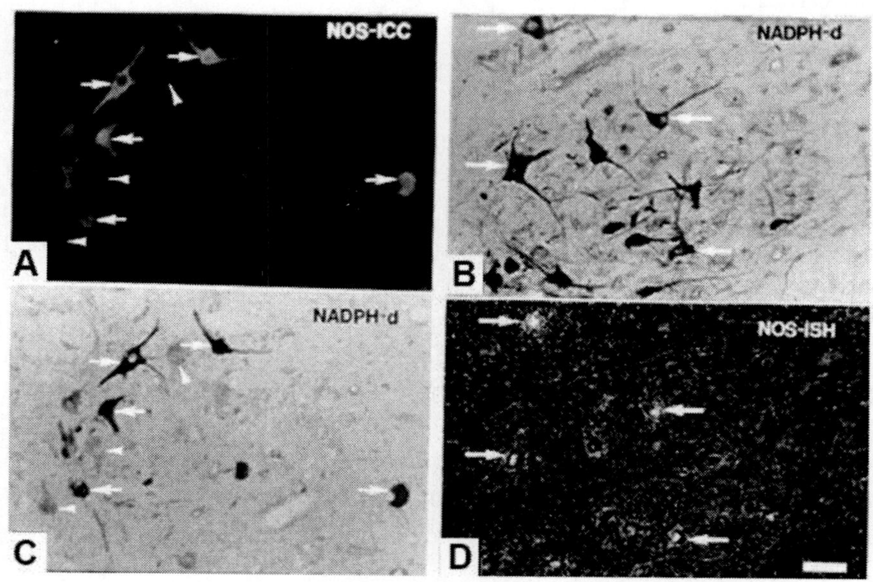

Figure 5 cross sections of ventral spinal cord showing NOS expression at 2 weeks after spinal root avulsion. Using the same spinal cord section, images (A) and (C) demonstrate that the same cell population were co-stained by nNOS immunostaining (A) and NADPH-d histochemistry (C). Arrows indicate the cells with dark staining while arrowheads represent light stained cells. In image (B) and (D), the section was firstly stained with NADPH-d histochemistry (B) followed with nNOS mRNA in situ hybridization. Arrows in the two photos show co-stained cells (Scale bar, 50 μm).

3.3.1.3. Rescue of Motoneuron by Application of NOS Inhibitors

The toxicity of high NO level, together with the coincidence of NOS expression and the motoneurons death after avulsion inspired the concept of promoting motoneuron survival by NOS inhibitors. Indeed, daily application of nitroarginine, a NOS inhibitor, into rats with C7 root avulsion increased the surviving motoneurons from 20% in control group to 60%, as observed at 6 weeks post injury [94]. This idea was further confirmed by the evidences that implantation of PN graft suppressed the NOS up-regulation after avulsion and eventually increased surviving motoneurons [60]. In addition, in cultured motoneurons, withdrawal of BDNF caused neuron death and treatment with

Nitro-L-arginine (L-NA) and nitro-L-arginine methyl ester (L-NAME) (NOS inhibitors) prevented motoneuron death [90]. However, there was also report that blocking NOS by antisense oligos actually deteriorate motoneuron death after avulsion injury [95]. This might imply that the toxicity of NO is dosage-dependent and basic level of NO is beneficial to neurons.

3.3.2. Inhibition of Caspase Cascade to Enhance Neuron Survival

3.3.2.1. Caspase-Dependent Cell Apoptosis

Caspases (cysteinyl aspartate proteinases) are a class of proteolytic enzymes, which cleave their substrate following an aspartate residue. They are evolutionarily conserved, with homologues found in *C. elegance, drosophila* and mammals [96, 97]. Their traditional function lies in their roles in apoptosis, as signal transmitters or executors of the machinery [96, 97]. In mammals, they also participant in many non-apoptosis processes, including cell survival, proliferation and differentiation [98, 99]. It is well established that, in mammals, there are two different mechanisms mediating caspase-dependent apoptosis, intrinsic and extrinsic pathways. In the former pathway, Bcl-2 family plays a major role in transforming the apoptosis stimuli and triggering mitochondrial outer membrane permeabilization (MOMP), which is a key event in the intrinsic apoptosis. The leakage of cytochrome c from mitochondria induces the formation of apoptosome. Then, initiator caspase-9 is activated by the apoptosome, followed with the activation of effector caspase-3 and/or caspase-7. In the extrinsic pathway, cell surface receptors, such as Fas, receive extracellular signals and recruit intracellular adaptor protein to form a death receptor-induced signaling complex (DISC). The complex further recruits and activates initiator caspase-8, which would in turn activate downstream effector caspases. In some cases, these two pathways coalesce. For example, caspase-8 could mediate the cleavage of Bid (Bcl-2 family member) and further induces the MOMP [99].

3.3.2.2. Caspase Inhibitors

The discovery of caspase inhibitor makes it possible to control and prevent cell death by manipulation of caspase cascade. The first identified caspase inhibitor was cytokine response modifier A (CrmA) from cowpox virus [100]. It was validated that CrmA was able to prevent mammalian cell death induced by overexpression of caspase-1 [101]. Since then, caspase inhibitors used to promote cell survival have expanded from viral to cellular and chemical products [102, 103]. P35, similar to CrmA, is another inhibitor originated from virus [104, 105]. They both bind to the catalytic domain of caspase as competitive pseudosubstrates. In addition, there is a long list of synthetic peptide caspase inhibitors, with variations on binding specificity and affinity [102]. Generally, they interact with the catalytic domain of caspase, and hence, are also competitive inhibitors. Yet, their capabilities of penetrating into cells, problems as proteolysis and toxicity are matters of concerns. Owning to these, compound inhibitors are developed, which target the caspase and prevent the conformational change to their active allosteric. For example, Compound 34 by Sunesis was found to be able to block the dimerization and activation of caspase-9 [106].

3.3.2.3. Inhibition of Caspase Cascade to Promote Neuron Survival

As a major player in apoptosis-mediated cell death, caspase has been considered as a potential target for cell fate reversion. Neurons are not out of expectations. Indeed, neurons share the similar apoptosis mechanisms as described in section 3.3.2.1. Caspase cascade is indicated in various types of neuron death, triggered by neurodegenerative diseases, neuron trauma and inflammation [107-110]. After traumatic brain injury, caspase-3 was elevated [108] and treatment with its inhibitor not only reduced the trauma area, but also decreased neuron loss [111]. After a spinal root avulsion injury, caspase inhibitors, either specifically binding to caspase-3 or without subtypes specificity, enabled over 70% of the injured motoneurons to stay alive at 3 weeks post injury [112]. However, this significant rescue effect was only observed in neonatal rats, but not in adult animals. Others tried to transduce virus encoding Bcl-2 inhibitor into motoneurons after avulsion. At 2 weeks after avulsion, the surviving motoneurons were up to 80% in Bcl-2 inhibited group, showing a protective effect comparable to that of GDNF [113].

3.3.3. Motoneuron Survival and Axonal Regeneration

Following the avulsion, the lesioned CNS-PNS transitional zone (TZ) undergoes rapid changes, including the formation of glial scar, which is similar to that found after a spinal cord injury [2, 71, 114]. Motoneuron survival and efficient axonal regeneration are both required for the functional recovery. The glial scar forms a barrier to regeneration. Chondroitin sulfate proteoglycan (CSPG) has been indicated as the major inhibitor for decades [115-117], but the mechanism of inhibition only started to get clear after the identification of its neuronal receptor, protein tyrosine phosphatase-σ (PTP-σ) [118]. Lately, it's been demonstrated that modulation of PTP-σ by a small peptide named intracellular sigma peptide (ISP) relieved the CSPG mediated inhibition [119] and remarkably promoted axonal regeneration after avulsion+ reimplantation [120]. Importantly, ISP treatment also increased surviving motoneurons from 62%, in control animals with avulsion+reimplantation, to 81% at 12 weeks post surgery, reflecting neuroprotection. Although the underline mechanism on ISP's capability of neuroprotection still remains to be elucidated, it shows therapeutic potential, because motoneuron survival and axonal regeneration were promoted simultaneously. Axonal regeneration might not only be essential for reconstruction of neuronal circuit, but also enable motoneurons to access to sufficient trophic factors provided by peripheral nerve and distal muscles.

3.3.4. Crosstalk between Different Strategies and Problems Remained

The spinal root avulsion injury, together with the following pathological changes, such as neurodegeneration, inflammation and regeneration, is complicated and cannot be regarded as an isolated process. A lot of cell types, including astrocytes, Schwann cells, microglia, macrophages and even distal muscle fibers can be involved, not to mention the lesioned motoneurons. The above-described strategies imply various responses and adjustments of different cell types to the avulsion injury. For instance, the disconnection with distal Schwann cells and target muscles deprives motoneurons of trophic factors, hence implantation of peripheral nerves or application of neurotrophic factors showed some extent of rescue. Apart from expression of nNOS by motoneurons themselves, iNOS from inflammation also produces and releases NO, which might participate in the caspase-involved apoptosis. Administration of BDNF and GDNF both eliminated NOS expression in avulsed motoneurons [24]. Astrocytes in the spinal cord synthesize inhibitory CSPG in the glial scar and suppress axonal regeneration after avulsion. Elimination of inhibitors in the scar tissue not only enhanced axonal regeneration, but also demonstrated neuroprotection [120]. Therefore, it's not surprising that different strategies have been combined for synergistic enhancement of motoneuron survival. Implantation of peripheral nerve with BDNF or GDNF treatment exhibited higher efficiency in neuroprotection than that of single invention alone [30]. NOS inhibitor and BNDF treatment in combination with PN graft highly improved the neuron survival and reduced the NOS immunoreactivity at the same time [121].

In clinic, the spinal root avulsion injury is devastating, leading to severe motor dysfunction. It's difficult to diagnose an avulsion injury immediately after the injury, whereas early treatment is demanded to promote the highest survival rate of motoneurons. Although neurotrophic factors are potential in the treatment, the dosage and application method should be considered with discretion, because high level has been indicated to lead to neuroma-like structure [122]. Caspase-mediated apoptosis is not the only pathway leading to dismantling motoneuron death, and caspases take part in other cellular processes apart from apoptosis. These issues not only remind us to discover more specific drugs, but also urge us to work on a clearer picture of avulsion-induced neuron death.

4. ASSESSMENT OF SURVIVING MOTONEURONS AFTER AVULSION INJURY

When studying avulsion injury in laboratory, it always requires us to assess the number of surviving motoneurons in the injured spinal cord. Nevertheless, it's not a simple work. As the avulsion injury induces changes of expression profile and cellular morphology, choosing the marker or dye carefully and setting the appropriate criterion would make a more precise measurement. Here, we briefly explain some commonly used techniques (Figure 6).

4.1. Nissl Staining

As early as 1886, Ehrlich utilized methylene blue as a neuron dye [123]. This method was quickly adopted and adjusted by Nissl in the staining of sections from an alcohol-hardened brain [124]. Basically, Nissl staining applies basic dyes to interact with negatively charged nucleic acids. Therefore, the rough endoplasmic reticulum (Nissl granules) of neurons can be darkly stained, due to the high amount of ribosomal RNA. Besides, there are also 1-2 visible nucleoli in the center of nucleus due to the condensed DNA. (Figure 6A) This method has been of importance in exhibiting cellular architecture, under normal or pathological circumstances. A wide range of dyes can be utilized, including thionin [125], neutral red and cresyl violet [126], and the color changes accordingly. As motoneuron is not the only cell type stained by Nissl method, it's important to single out motoneurons by proper morphological criterions [127]. In practice, motoneurons are supposed to have a large nucleus with 1-2 visible nucleoli and an apparently large cytoplasm (the minimum diameters used ranging from 25-35um) [50, 128].

4.2. ChAT Immunostainning

Choline acetyltransferase (ChAT) is the enzyme transferring acetyl group to choline to produce the neurotransmitter acetylcholine. It is presented in cholinergic neurons. In spinal motoneurons, the expression of ChAT can be detected at both protein and mRNA levels, in neonatal and adult animals [129-131]. Following axotomy, there exists a down-regulation of ChAT and the extent of decrease depends on the type of injury [17, 132]. Spinal root avulsion induced attenuated ChAT level in at least some subtypes of the lesioned motoneurons, within the first week of injury. However, the protein quantity goes back to detectable level in surviving motoneurons at 4 weeks post avulsion [17]. Therefore, ChAT inmmunohistochemistry can be used to quantify the number of motoneurons in normal animals or those with long-term injuries [133, 134] (Figure 6B).

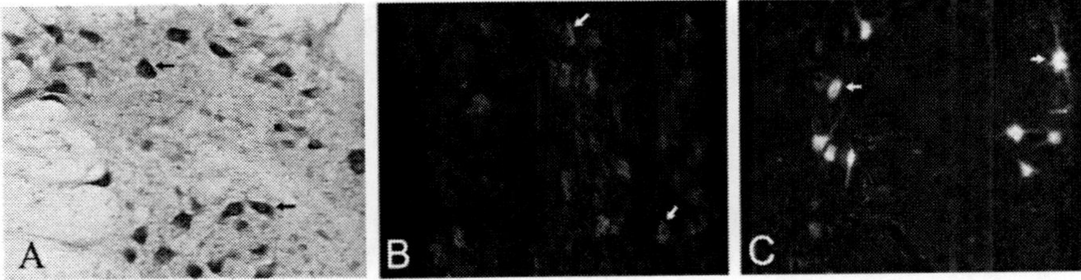

Figure 6. Results from different staining methods to display motoneurons for measurement. (A) Nissl staining by neutral red; (B) ChAT immunohistochemistry; (C) Fluorogold labeling. Arrows indicate labeled motoneurons.

4.3. Retrograde Labeling

When avulsion is followed with implantation of peripheral nerve or re-implantation of avulsed spinal roots, axonal regeneration occurs. Retrograde labeling can be adopted to quantify surviving and regenerating motoneurons. Fluorogold is a commonly used dye [120]. Once absorbed by axons, it can be retrogradely transported to soma and dendrites (Figure 6C). Via injecting Fluorogold into the implanted nerve, it stains all the motoneurons, which have extended their axons to the injecting site. The technique should be carefully performed, to prevent the leakage, and to ensure that all the axons have taken in the dye.

REFERENCES

[1] Carlstedt T. Central Nerve Plexus Injury: Imperial College Press; 2007.

[2] Carlstedt T. Root repair review: basic science background and clinical outcome. *Restor Neurol Neurosci.* 2008;26(2-3):225-41.

[3] Leffert RD. Brachial plexus injuries: Churchill Livingstone; 1985.

[4] Koliatsos VE, Price WL, Pardo CA, Price DL. Ventral root avulsion: an experimental model of death of adult motor neurons. *The Journal of comparative neurology.* 1994;342(1):35-44.

[5] Olsson T, Lundberg C, Lidman O, Piehl F. Genetic regulation of nerve avulsion-induced spinal cord inflammation. *Ann N Y Acad Sci.* 2000;917:186-96.

[6] Li L, Wu W, Lin LF, Lei M, Oppenheim RW, Houenou LJ. Rescue of adult mouse motoneurons from injury-induced cell death by glial cell line-derived neurotrophic factor. *Proceedings of the National Academy of Sciences of the United States of America.* 1995;92(21):9771-5.

[7] Wu W. Expression of nitric-oxide synthase (NOS) in injured CNS neurons as shown by NADPH diaphorase histochemistry. *Experimental neurology.* 1993;120(2):153-9.

[8] Gu Y, Spasic Z, Wu W. The effects of remaining axons on motoneuron survival and NOS expression following axotomy in the adult rat. *Developmental neuroscience.* 1997;19(3):255-9.

[9] Taniuchi M, Clark HB, Schweitzer JB, Johnson EM, Jr. Expression of nerve growth factor receptors by Schwann cells of axotomized peripheral nerves: ultrastructural location, suppression by axonal contact, and binding properties. *The Journal of neuroscience: the official journal of the Society for Neuroscience.* 1988;8(2):664-81.

[10] Frostick SP, Yin Q, Kemp GJ. Schwann cells, neurotrophic factors, and peripheral nerve regeneration. *Microsurgery.* 1998;18(7):397-405.

[11] Terenghi G. Peripheral nerve regeneration and neurotrophic factors. *J Anat. 1999*;194 (Pt 1):1-14.

[12] Boyd JG, Gordon T. Neurotrophic factors and their receptors in axonal regeneration and functional recovery after peripheral nerve injury. *Mol Neurobiol.* 2003;27(3):277-324.

[13] Madduri S, Gander B. Schwann cell delivery of neurotrophic factors for peripheral nerve regeneration. *J Peripher Nerv Syst.* 2010;15(2):93-103.

[14] Martin LJ, Kaiser A, Price AC. Motor neuron degeneration after sciatic nerve avulsion in adult rat evolves with oxidative stress and is apoptosis. *J Neurobiol.* 1999;40(2):185-201.

[15] Park OH, Lee KJ, Rhyu IJ, Geum D, Kim H, Buss R, et al. Bax-dependent and -independent death of motoneurons after facial nerve injury in adult mice. *The European journal of neuroscience.* 2007;26(6):1421-32.

[16] Ohlsson M, Havton LA. Complement activation after lumbosacral ventral root avulsion injury. *Neuroscience letters.* 2006;394(3):179-83.

[17] Hoang TX, Nieto JH, Tillakaratne NJ, Havton LA. Autonomic and motor neuron death is progressive and parallel in a lumbosacral ventral root avulsion model of cauda equina injury. *The Journal of comparative neurology.* 2003;467(4):477-86.

[18] Li L, Houenou LJ, Wu W, Lei M, Prevette DM, Oppenheim RW. Characterization of spinal motoneuron degeneration following different types of peripheral nerve injury in neonatal and adult mice. *The Journal of comparative neurology.* 1998;396(2):158-68.

[19] Martin LJ, Liu Z. Injury-induced spinal motor neuron apoptosis is preceded by DNA single-strand breaks and is p53- and Bax-dependent. *J Neurobiol.* 2002;50(3):181-97.

[20] Martin LJ, Kaiser A, Yu JW, Natale JE, Al-Abdulla NA. Injury-induced apoptosis of neurons in adult brain is mediated by p53-dependent and p53-independent pathways and requires Bax. *The Journal of comparative neurology.* 2001;433(3):299-311.

[21] Al-Abdulla NA, Martin LJ. Apoptosis of retrogradely degenerating neurons occurs in association with the accumulation of perikaryal mitochondria and oxidative damage to the nucleus. *Am J Pathol.* 1998;153(2):447-56.

[22] Bellander BM, von Holst H, Fredman P, Svensson M. Activation of the complement cascade and increase of clusterin in the brain following a cortical contusion in the adult rat. *Journal of neurosurgery.* 1996;85(3):468-75.

[23] Yang Y, Xie Y, Chai H, Fan M, Liu S, Liu H, et al. Microarray analysis of gene expression patterns in adult spinal motoneurons after different types of axonal injuries. *Brain research.* 2006;1075(1):1-12.

[24] Wu W, Li L, Yick LW, Chai H, Xie Y, Yang Y, et al. GDNF and BDNF alter the expression of neuronal NOS, c-Jun, and p75 and prevent motoneuron death following spinal root avulsion in adult rats. *Journal of neurotrauma.* 2003;20(6):603-12.

[25] Zhao S, Pang Y, Beuerman RW, Thompson HW, Kline DG. Expression of c-Fos protein in the spinal cord after brachial plexus injury: comparison of root avulsion and distal nerve transection. *Neurosurgery.* 1998;42(6):1357-62; discussion 62-3.

[26] Yuan Q, Hu B, Su H, So KF, Lin Z, Wu W. GAP-43 expression correlates with spinal motoneuron regeneration following root avulsion. *J Brachial Plex Peripher Nerve Inj.* 2009;4:18.

[27] Fu Y, Hashimoto M, Ino H, Murakami M, Yamazaki M, Moriya H. Spinal root avulsion-induced upregulation of osteopontin expression in the adult rat spinal cord. *Acta neuropathologica.* 2004;107(1):8-16.

[28] Yuan Q, Hu B, Chu TH, Su H, Zhang W, So KF, et al. Co-expression of GAP-43 and nNOS in avulsed motoneurons and their potential role for motoneuron regeneration. *Nitric Oxide.* 2010;23(4):258-63.

[29] Barde YA. What, if anything, is a neurotrophic factor? *Trends Neurosci.* 1988;11(8):343-6.

[30] Chu TH, Li SY, Guo A, Wong WM, Yuan Q, Wu W. Implantation of neurotrophic factor-treated sensory nerve graft enhances survival and axonal regeneration of motoneurons after spinal root avulsion. *Journal of neuropathology and experimental neurology.* 2009;68(1):94-101.

[31] Levi-Montalcini R, Hamburger V. Selective growth stimulating effects of mouse sarcoma on the sensory and sympathetic nervous system of the chick embryo. *J Exp Zool.* 1951;116(2):321-61.

[32] Cohen S. Purification of a Nerve-Growth Promoting Protein from the Mouse Salivary Gland and Its Neuro-Cytotoxic Antiserum. *Proceedings of the National Academy of Sciences of the United States of America.* 1960;46(3):302-11.

[33] Hofer MM, Barde YA. Brain-derived neurotrophic factor prevents neuronal death in vivo. *Nature.* 1988;331(6153):261-2.

[34] Hohn A, Leibrock J, Bailey K, Barde YA. Identification and characterization of a novel member of the nerve growth factor/brain-derived neurotrophic factor family. *Nature.* 1990;344(6264):339-41.

[35] Maisonpierre PC, Belluscio L, Squinto S, Ip NY, Furth ME, Lindsay RM, et al. Neurotrophin-3: a neurotrophic factor related to NGF and BDNF. *Science.* 1990;247(4949 Pt 1):1446-51.

[36] Lin LF, Doherty DH, Lile JD, Bektesh S, Collins F. GDNF: a glial cell line-derived neurotrophic factor for midbrain dopaminergic neurons. *Science.* 1993;260(5111):1130-2.

[37] Adler R, Landa KB, Manthorpe M, Varon S. Cholinergic neuronotrophic factors: intraocular distribution of trophic activity for ciliary neurons. *Science.* 1979;204(4400):1434-6.

[38] Chao MV. Neurotrophins and their receptors: a convergence point for many signalling pathways. *Nat Rev Neurosci.* 2003;4(4):299-309.

[39] Lu B, Pang PT, Woo NH. The yin and yang of neurotrophin action. *Nat Rev Neurosci.* 2005;6(8):603-14.

[40] Cohen S, McElroy W, Glass B. Chemical Basis of Development. ed McElroy, WD, and. 1958:665-77.

[41] Henderson JE, Amizuka N, Warshawsky H, Biasotto D, Lanske BM, Goltzman D, et al. Nucleolar localization of parathyroid hormone-related peptide enhances survival of chondrocytes under conditions that promote apoptotic cell death. *Mol Cell Biol.* 1995;15(8):4064-75.

[42] Zurn AD, Baetge EE, Hammang JP, Tan SA, Aebischer P. Glial cell line-derived neurotrophic factor (GDNF), a new neurotrophic factor for motoneurones. Neuroreport. 1994;6(1):113-8.

[43] Gill SS, Patel NK, Hotton GR, O'Sullivan K, McCarter R, Bunnage M, et al. Direct brain infusion of glial cell line-derived neurotrophic factor in Parkinson disease. *Nat Med.* 2003;9(5):589-95.

[44] Springer JE, Mu X, Bergmann LW, Trojanowski JQ. Expression of GDNF mRNA in rat and human nervous tissue. *Experimental neurology.* 1994;127(2):167-70.

[45] Lie DC, Weis J. GDNF expression is increased in denervated human skeletal muscle. *Neuroscience letters.* 1998;250(2):87-90.

[46] Barde YA, Edgar D, Thoenen H. Purification of a new neurotrophic factor from mammalian brain. *EMBO J.* 1982;1(5):549-53.

[47] Novikov L, Novikova L, Kellerth JO. Brain-derived neurotrophic factor promotes survival and blocks nitric oxide synthase expression in adult rat spinal motoneurons after ventral root avulsion. *Neuroscience letters.* 1995;200(1):45-8.

[48] Kishino A, Ishige Y, Tatsuno T, Nakayama C, Noguchi H. BDNF prevents and reverses adult rat motor neuron degeneration and induces axonal outgrowth. *Experimental neurology.* 1997;144(2):273-86.

[49] Chai H, Wu W, So KF, Prevette DM, Oppenheim RW. Long-term effects of a single dose of brain-derived neurotrophic factor on motoneuron survival following spinal root avulsion in the adult rat. *Neuroscience letters.* 1999;274(3):147-50.

[50] Buck CR, Seburn KL, Cope TC. Neurotrophin expression by spinal motoneurons in adult and developing rats. *The Journal of comparative neurology.* 2000;416(3):309-18.

[51] Al-Majed AA, Brushart TM, Gordon T. Electrical stimulation accelerates and increases expression of BDNF and trkB mRNA in regenerating rat femoral motoneurons. *The European journal of neuroscience.* 2000;12(12):4381-90.

[52] Jiang M, Peng Q, Liu X, Jin J, Hou Z, Zhang J, et al. Small-molecule TrkB receptor agonists improve motor function and extend survival in a mouse model of Huntington's disease. *Hum Mol Genet.* 2013;22(12):2462-70.

[53] Jang SW, Liu X, Yepes M, Shepherd KR, Miller GW, Liu Y, et al. A selective TrkB agonist with potent neurotrophic activities by 7,8-dihydroxyflavone. *Proceedings of the National Academy of Sciences of the United States of America.* 2010;107(6):2687-92.

[54] Shu XQ, Llinas A, Mendell LM. Effects of trkB and trkC neurotrophin receptor agonists on thermal nociception: a behavioral and electrophysiological study. *Pain.* 1999;80(3):463-70.

[55] Zhou LH, Wu W. Survival of injured spinal motoneurons in adult rat upon treatment with glial cell line-derived neurotrophic factor at 2 weeks but not at 4 weeks after root avulsion. *Journal of neurotrauma.* 2006;23(6):920-7.

[56] Wu W, Liuzzi FJ, Schinco FP, Depto AS, Li Y, Mong JA, et al. Neuronal nitric oxide synthase is induced in spinal neurons by traumatic injury. *Neuroscience.* 1994;61(4):719-26.

[57] Campbell WW. Evaluation and management of peripheral nerve injury. Clinical neurophysiology: official journal of the International *Federation of Clinical Neurophysiology.* 2008;119(9):1951-65.

[58] Khuong HT, Midha R. Advances in nerve repair. *Curr Neurol Neurosci Rep.* 2013;13(1):322.

[59] Horvat JC, Pecot-Dechavassine M, Mira JC, Davarpanah Y. Formation of functional endplates by spinal axons regenerating through a peripheral nerve graft. A study in the adult rat. *Brain research bulletin.* 1989;22(1):103-14.

[60] Wu W, Han K, Li L, Schinco FP. Implantation of PNS graft inhibits the induction of neuronal nitric oxide synthase and enhances the survival of spinal motoneurons following root avulsion. *Experimental neurology.* 1994;129(2):335-9.

[61] Wu W, Chai H, Zhang J, Gu H, Xie Y, Zhou L. Delayed implantation of a peripheral nerve graft reduces motoneuron survival but does not affect regeneration following spinal root avulsion in adult rats. *Journal of neurotrauma.* 2004;21(8):1050-8.

[62] Holtzer CA, Marani E, van Dijk GJ, Thomeer RT. Repair of ventral root avulsion using autologous nerve grafts in cats. *J Peripher Nerv Syst.* 2003;8(1):17-22.

[63] Bertelli JA, Mira JC. Brachial plexus repair by peripheral nerve grafts directly into the spinal cord in rats. Behavioral and anatomical evidence of functional recovery. *Journal of neurosurgery.* 1994;81(1):107-14.

[64] Terzis JK, Liberson WT, Peffley C, Carson KA, Spurrier D. Reinnervation of peripheral nerve grafts by spinal cord fibres tested by transcranial brain stimulation. *Electromyogr Clin Neurophysiol.* 1989;29(7-8):417-23.

[65] Chu TH, Du Y, Wu W. Motor nerve graft is better than sensory nerve graft for survival and regeneration of motoneurons after spinal root avulsion in adult rats. *Experimental neurology.* 2008;212(2):562-5.

[66] Hoffmann CF, Marani E, van Dijk JG, vd Kamp W, Thomeer RT. Reinnervation of avulsed and reimplanted ventral rootlets in the cervical spinal cord of the cat. *Journal of neurosurgery.* 1996;84(2):234-43.

[67] Hoffmann CF, Thomeer RT, Marani E. Reimplantation of ventral rootlets into the cervical spinal cord after their avulsion: an anterior surgical approach. *Clin Neurol Neurosurg.* 1993;95 Suppl:S112-8.

[68] Chai H, Wu W, So KF, Yip HK. Survival and regeneration of motoneurons in adult rats by reimplantation of ventral root following spinal root avulsion. *Neuroreport.* 2000;11(6):1249-52.

[69] Fournier HD, Menei P, Khalifa R, Mercier P. Ideal intraspinal implantation site for the repair of ventral root avulsion after brachial plexus injury in humans. A preliminary anatomical study. *Surg Radiol Anat.* 2001;23(3):191-5.

[70] Smith KJ, Kodama RT. Reinnervation of denervated skeletal muscle by central neurons regenerating via ventral roots implanted into the spinal cord. *Brain research.* 1991;551(1-2):221-9.

[71] Eggers R, Tannemaat MR, Ehlert EM, Verhaagen J. A spatio-temporal analysis of motoneuron survival, axonal regeneration and neurotrophic factor expression after lumbar ventral root avulsion and implantation. *Exp Neurol.* 2010;223(1):207-20.

[72] Su H, Yuan Q, Qin D, Yang X, Wong WM, So KF, et al. Ventral root re-implantation is better than peripheral nerve transplantation for motoneuron survival and regeneration after spinal root avulsion injury. *BMC surgery*. 2013;13(1):21.

[73] Carlstedt T, Aldskogius H, Hallin RG, Nilsson-Remahl I. Novel surgical strategies to correct neural deficits following experimental spinal nerve root lesions. *Brain research bulletin*. 1993;30(3-4):447-51.

[74] Carlstedt T, Linda H, Cullheim S, Risling M. Reinnervation of hind limb muscles after ventral root avulsion and implantation in the lumbar spinal cord of the adult rat. *Acta Physiol Scand*. 1986;128(4):645-6.

[75] Gu HY, Chai H, Zhang JY, Yao ZB, Zhou LH, Wong WM, et al. Survival, regeneration and functional recovery of motoneurons in adult rats by reimplantation of ventral root following spinal root avulsion. *The European journal of neuroscience*. 2004;19(8):2123-31.

[76] Gu HY, Chai H, Zhang JY, Yao ZB, Zhou LH, Wong WM, et al. Survival, regeneration and functional recovery of motoneurons after delayed reimplantation of avulsed spinal root in adult rat. *Exp Neurol*. 2005;192(1):89-99.

[77] Moissonnier P, Duchossoy Y, Lavieille S, Horvat JC. Lateral approach of the dog brachial plexus for ventral root reimplantation. *Spinal Cord*. 1998;36(6):391-8.

[78] Risling M, Cullheim S, Hildebrand C. Reinnervation of the ventral root L7 from ventral horn neurons following intramedullary axotomy in adult cats. *Brain research*. 1983;280(1):15-23.

[79] Cullheim S, Carlstedt T, Linda H, Risling M, Ulfhake B. Motoneurons reinnervate skeletal muscle after ventral root implantation into the spinal cord of the cat. *Neuroscience*. 1989;29(3):725-33.

[80] Hallin RG, Carlstedt T, Nilsson-Remahl I, Risling M. Spinal cord implantation of avulsed ventral roots in primates; correlation between restored motor function and morphology. *Experimental brain research Experimentelle Hirnforschung Experimentation cerebrale*. 1999;124(3):304-10.

[81] Carlstedt T, Grane P, Hallin RG, Noren G. Return of function after spinal cord implantation of avulsed spinal nerve roots. *Lancet*. 1995;346(8986):1323-5.

[82] Furchgott RF, Zawadzki JV. The obligatory role of endothelial cells in the relaxation of arterial smooth muscle by acetylcholine. *Nature*. 1980;288(5789):373-6.

[83] Bult H, Boeckxstaens GE, Pelckmans PA, Jordaens FH, Van Maercke YM, Herman AG. Nitric oxide as an inhibitory non-adrenergic non-cholinergic neurotransmitter. *Nature.* 1990;345(6273):346-7.

[84] Bredt DS, Snyder SH. Nitric oxide, a novel neuronal messenger. *Neuron.* 1992;8(1):3-11.

[85] Boeckxstaens GE, Pelckmans PA, Bult H, De Man JG, Herman AG, van Maercke YM. Evidence for nitric oxide as mediator of non-adrenergic non-cholinergic relaxations induced by ATP and GABA in the canine gut. *Br J Pharmacol.* 1991;102(2):434-8.

[86] Dawson TM, Dawson VL, Snyder SH. A novel neuronal messenger molecule in brain: the free radical, nitric oxide. *Ann Neurol.* 1992;32(3):297-311.

[87] Bossy-Wetzel E, Talantova MV, Lee WD, Scholzke MN, Harrop A, Mathews E, et al. Crosstalk between nitric oxide and zinc pathways to neuronal cell death involving mitochondrial dysfunction and p38-activated K+ channels. *Neuron.* 2004;41(3):351-65.

[88] Bal-Price A, Brown GC. Inflammatory neurodegeneration mediated by nitric oxide from activated glia-inhibiting neuronal respiration, causing glutamate release and excitotoxicity. *The Journal of neuroscience: the official journal of the Society for Neuroscience.* 2001;21(17):6480-91.

[89] Brown GC. Nitric oxide and neuronal death. *Nitric Oxide.* 2010;23(3):153-65.

[90] Estevez AG, Spear N, Manuel SM, Radi R, Henderson CE, Barbeito L, et al. Nitric oxide and superoxide contribute to motor neuron apoptosis induced by trophic factor deprivation. *The Journal of neuroscience: the official journal of the Society for Neuroscience.* 1998;18(3):923-31.

[91] Forstermann U, Sessa WC. Nitric oxide synthases: regulation and function. *Eur Heart J.* 2012;33(7):829-37, 37a-37d.

[92] Wu W. Response of nitric oxide snythase to neuronal injury. *Handbook of Chemical Neuroanatomy Vol 17: Functional Neuroanatomy of the Nitric Oxide System.* 2000.

[93] Li Y, Wu W, Schinco F, Goode G, editors. Sciatic nerve transection causes expression of nitric oxide synthase (NOS) and death of spinal motoneurons in newborn and earlier postnatal rats. *Soc Neurosci Abstr;* 1993.

[94] Wu W, Li L. Inhibition of nitric oxide synthase reduces motoneuron death due to spinal root avulsion. *Neuroscience letters.* 1993;153(2):121-4.

[95] Zhou L, Wu W. Antisense oligos to neuronal nitric oxide synthase aggravate motoneuron death induced by spinal root avulsion in adult rat. *Experimental neurology*. 2006;197(1):84-92.

[96] Kumar S. Caspase function in programmed cell death. *Cell Death Differ*. 2007;14(1):32-43.

[97] Kroemer G, Martin SJ. Caspase-independent cell death. *Nat Med*. 2005;11(7):725-30.

[98] Lamkanfi M, Festjens N, Declercq W, Vanden Berghe T, Vandenabeele P. Caspases in cell survival, proliferation and differentiation. *Cell Death Differ*. 2007;14(1):44-55.

[99] D'Amelio M, Cavallucci V, Cecconi F. Neuronal caspase-3 signaling: not only cell death. *Cell Death Differ*. 2010;17(7):1104-14.

[100] Ray CA, Black RA, Kronheim SR, Greenstreet TA, Sleath PR, Salvesen GS, et al. Viral inhibition of inflammation: cowpox virus encodes an inhibitor of the interleukin-1 beta converting enzyme. *Cell*. 1992;69(4):597-604.

[101] Miura M, Zhu H, Rotello R, Hartwieg EA, Yuan J. Induction of apoptosis in fibroblasts by IL-1 beta-converting enzyme, a mammalian homolog of the C. elegans cell death gene ced-3. *Cell*. 1993;75(4):653-60.

[102] Callus BA, Vaux DL. Caspase inhibitors: viral, cellular and chemical. *Cell Death Differ*. 2007;14(1):73-8.

[103] Ekert PG, Silke J, Vaux DL. Caspase inhibitors. *Cell Death Differ*. 1999;6(11):1081-6.

[104] Clem RJ, Fechheimer M, Miller LK. Prevention of apoptosis by a baculovirus gene during infection of insect cells. *Science*. 1991;254(5036):1388-90.

[105] LaCount DJ, Hanson SF, Schneider CL, Friesen PD. Caspase inhibitor P35 and inhibitor of apoptosis Op-IAP block in vivo proteolytic activation of an effector caspase at different steps. *J Biol Chem*. 2000;275(21):15657-64.

[106] Scheer JM, Romanowski MJ, Wells JA. A common allosteric site and mechanism in caspases. *Proceedings of the National Academy of Sciences of the United States of America*. 2006;103(20):7595-600.

[107] Troy CM, Rabacchi SA, Friedman WJ, Frappier TF, Brown K, Shelanski ML. Caspase-2 mediates neuronal cell death induced by beta-amyloid. *The Journal of neuroscience: the official journal of the Society for Neuroscience*. 2000;20(4):1386-92.

[108] Yakovlev AG, Knoblach SM, Fan L, Fox GB, Goodnight R, Faden AI. Activation of CPP32-like caspases contributes to neuronal apoptosis and neurological dysfunction after traumatic brain injury. *The Journal of neuroscience: the official journal of the Society for Neuroscience.* 1997;17(19):7415-24.

[109] Eldadah BA, Faden AI. Caspase pathways, neuronal apoptosis, and CNS injury. *Journal of neurotrauma.* 2000;17(10):811-29.

[110] Burguillos MA, Deierborg T, Kavanagh E, Persson A, Hajji N, Garcia-Quintanilla A, et al. Caspase signalling controls microglia activation and neurotoxicity. *Nature.* 2011;472(7343):319-24.

[111] Clark RS, Kochanek PM, Watkins SC, Chen M, Dixon CE, Seidberg NA, et al. Caspase-3 mediated neuronal death after traumatic brain injury in rats. *J Neurochem.* 2000;74(2):740-53.

[112] Chan YM, Wu W, Yip HK, So KF, Oppenheim RW. Caspase inhibitors promote the survival of avulsed spinal motoneurons in neonatal rats. *Neuroreport.* 2001;12(3):541-5.

[113] Natsume A, Mata M, Wolfe D, Oligino T, Goss J, Huang S, et al. Bcl-2 and GDNF delivered by HSV-mediated gene transfer after spinal root avulsion provide a synergistic effect. *Journal of neurotrauma.* 2002;19(1):61-8.

[114] Fraher JP. The transitional zone and CNS regeneration. *J Anat.* 2000;196 (Pt 1):137-58.

[115] Bradbury EJ, Moon LD, Popat RJ, King VR, Bennett GS, Patel PN, et al. Chondroitinase ABC promotes functional recovery after spinal cord injury. *Nature.* 2002;416(6881):636-40.

[116] Cregg JM, DePaul MA, Filous AR, Lang BT, Tran A, Silver J. Functional regeneration beyond the glial scar. *Experimental neurology.* 2014;253:197-207.

[117] Yick LW, Wu W, So KF, Yip HK, Shum DK. Chondroitinase ABC promotes axonal regeneration of Clarke's neurons after spinal cord injury. *Neuroreport. 2000*;11(5):1063-7.

[118] Shen Y, Tenney AP, Busch SA, Horn KP, Cuascut FX, Liu K, et al. PTPsigma is a receptor for chondroitin sulfate proteoglycan, an inhibitor of neural regeneration. *Science.* 2009;326(5952):592-6.

[119] Lang BT, Cregg JM, DePaul MA, Tran AP, Xu K, Dyck SM, et al. Modulation of the proteoglycan receptor PTPsigma promotes recovery after spinal cord injury. *Nature.* 2015;518(7539):404-8.

[120] Li H, Wong C, Li W, Ruven C, He L, Wu X, et al. Enhanced regeneration and functional recovery after spinal root avulsion by manipulation of the proteoglycan receptor PTPsigma. *Sci Rep.* 2015;5:14923.

[121] Yick LW, Wu W, So KF. Additive effect of NOS inhibitor and neurotrophic factors on the survival of injured Clarke's neurons. *Neuroreport.* 1999;10(12):2569-73.

[122] Eggers R, Hendriks WT, Tannemaat MR, van Heerikhuize JJ, Pool CW, Carlstedt TP, et al. Neuroregenerative effects of lentiviral vector-mediated GDNF expression in reimplanted ventral roots. *Mol Cell Neurosci.* 2008;39(1):105-17.

[123] Windhorst U, Johansson H. Modern Techniques in Neuroscience Research: 33 Tables: Springer Science and Business Media; 1999.

[124] Nissl F. Über die sogenannten Granula der Nervenzellen: Neurol Zbl; 1894.

[125] Windle WF, Rhines R, Rankin J. A Nissl method using buffered solutions of thionin. *Biotechnic and Histochemistry.* 1943;18(2):77-86.

[126] Powers MM, Clark G. An evaluation of cresyl echt violet acetate as a Nissl stain. *Stain Technol.* 1955;30(2):83-8.

[127] Clarke P, Oppenheim R. Neuron Death in Vertebrate Development: In Vivo Methods. *Methods in cell biology.* 1995;46:277-321.

[128] Li L, Oppenheim RW, Lei M, Houenou LJ. Neurotrophic agents prevent motoneuron death following sciatic nerve section in the neonatal mouse. *J Neurobiol.* 1994;25(7):759-66.

[129] Lauterborn JC, Isackson PJ, Montalvo R, Gall CM. In situ hybridization localization of choline acetyltransferase mRNA in adult rat brain and spinal cord. *Brain Res Mol Brain Res.* 1993;17(1-2):59-69.

[130] Houser CR, Crawford GD, Barber RP, Salvaterra PM, Vaughn JE. Organization and morphological characteristics of cholinergic neurons: an immunocytochemical study with a monoclonal antibody to choline acetyltransferase. *Brain research.* 1983;266(1):97-119.

[131] Wetts R, Vaughn JE. Choline acetyltransferase and NADPH diaphorase are co-expressed in rat spinal cord neurons. *Neuroscience.* 1994;63(4):1117-24.

[132] Kou SY, Chiu AY, Patterson PH. Differential regulation of motor neuron survival and choline acetyltransferase expression following axotomy. *J Neurobiol.* 1995;27(4):561-72.

[133] Ohlsson M, Nieto JH, Christe KL, Havton LA. Long-term effects of a lumbosacral ventral root avulsion injury on axotomized motor neurons and avulsed ventral roots in a non-human primate model of cauda equina injury. *Neuroscience.* 2013;250:129-39.

[134] Blits B, Carlstedt TP, Ruitenberg MJ, de Winter F, Hermens WT, Dijkhuizen PA, et al. Rescue and sprouting of motoneurons following ventral root avulsion and reimplantation combined with intraspinal adeno-associated viral vector-mediated expression of glial cell line-derived neurotrophic factor or brain-derived neurotrophic factor. *Experimental neurology.* 2004;189(2):303-16.

In: Horizons in Neuroscience Research. Vol. 25   ISBN: 978-1-63485-286-9
Editors: A. Costa and E. Villalba   © 2016 Nova Science Publishers, Inc.

Chapter 5

SPATIAL PATTERNS OF PHOSPHORYLATED TDP-43-IMMUNOREACTIVE CELLULAR INCLUSIONS IN FAMILIAL AND SPORADIC FRONTOTEMPORAL LOBAR DEGENERATION WITH TDP-43 PROTEINOPATHY

Richard A. Armstrong[*]

Vision Sciences, Aston University, Birmingham, UK

ABSTRACT

Abnormal protein aggregates of transactive response (TAR) DNA-binding protein (TDP-43) in the form of neuronal cytoplasmic inclusions (NCI), oligodendroglial inclusions (GI), neuronal internuclear inclusions (NII), and dystrophic neurites (DN) are the pathological hallmark of frontotemporal lobar degeneration with TDP-43 proteinopathy (FTLD-TDP). To investigate the role of phosphorylated TDP-43 (pTDP-43) in neurodegeneration in FTLD-TDP, the spatial patterns of the pTDP-43-immunoreactive NCI, GI, NII, and DN were studied in frontal and temporal cortex in three groups of cases: (1) familial FTLD-TDP caused by *progranulin* (*GRN*) mutation, (2) a miscellaneous group of familial cases containing cases caused by *valosin-containing protein* (VCP)

---

[*] Corresponding Author: Dr. R.A. Armstrong, Vision Sciences, Aston University, Birmingham, B4 7ET, UK. Tel 0121-359-3611, Fax 0121-333-4220, EMail: R.A.Armstrong@aston.ac.uk

mutation, *ubiquitin associated protein 1 (UBAP1)* mutation, and cases not associated with currently known genes, and (3) sporadic FTLD-TDP. In a significant number of brain regions, the pTDP-43-immunoreactive inclusions developed in clusters and the clusters were distributed regularly parallel to the tissue boundary. The spatial patterns of the inclusions were similar to those revealed by a phosphorylation-independent anti-TDP-43 antibody. The spatial patterns and cluster sizes of the pTDP-43-immunoreactive inclusions were similar in *GRN* mutation cases, remaining familial cases, and in sporadic FTLD-TDP. Hence, pathological changes initiated by different genetic factors in familial cases and by unknown causes in sporadic FTLD-TDP appear to follow a parallel course resulting in very similar patterns of degeneration of frontal and temporal lobes.

Keywords: Frontotemporal lobar degeneration with TDP-43 proteinopathy (FTLD-TDP), TAR DNA-binding protein of 43 kDa (TDP-43), Neuronal cytoplasmic inclusions (NCI), Phosphorylation-dependent anti-TDP-43 antibody (pTDP-43), Spatial pattern

INTRODUCTION

The transactive response (TAR) DNA-binding protein of 43kDa (TDP-43) is an RNA binding protein encoded by the highly conserved *TARDPB* gene located on chromosome 1. TDP proteins have a glycine-rich domain and are believed to carry out essential cellular functions including the regulation of transcription, alternate splicing, and acting as a framework for nuclear bodies (Wang et al., 2004). TDP-43 is a nuclear protein containing two RNA-recognition sequences as well as a glycine-rich C-terminal sequence and is widely expressed in body tissues (Neumann et al., 2007). In neurodegenerative disease, TDP-43 may be redistributed from the nucleus to the cytoplasm, is ubiquinated, hyperphosphorylated, and then cleaved to generate toxic C-terminal fragments (Neumann et al., 2007). These fragments then accumulate to form cellular inclusions which may cause death of neurons and ultimately, neurodegeneration.

Recent studies suggest a significant role for TDP-43 in neurodegenerative disease. Hence, TDP-43 is a major pathological protein in frontotemporal lobar degeneration with TDP-43 proteinopathy (FTLD-TDP) previously called FTLD with ubiquitin immunoreactive inclusions (FTLD-U) (Cairns et al., 2007a), a disorder that accounts for approximately 10% of early-onset dementias. The spectrum of TDP proteinopathies includes most cases of sporadic and familial FTLD-TDP with or without associated motor neuron disease (MND). Accumulation of TDP-43 may also occur in Alzheimer's disease (AD) (Uryu et al., 2008), dementia with Lewy bodies (DLB) (Higashi et al., 2007), amylotrophic lateral sclerosis (ALS) (Brandmeir et al., 2008), hippocampal sclerosis (HS) (Amandir-Ortiz et al., 2007), and Guam-parkinsonism dementia complex (G-PDC) (Hasegawa et al., 2007).

FTLD-TDP is characterized by a widespread atrophy largely affecting the frontal, temporal, and parietal lobes with neuronal loss, microvacuolation in the superficial cortical laminae, reactive atrocytosis, and the presence of TDP-43-immunoreactive inclusions in both neurons and glial cells (Cairns et al., 2007b). The cellular inclusions include neuronal cytoplasmic inclusions (NCI), oligodendroglial inclusions (GI), neuronal internuclear inclusions (NII), and dystrophic neurites (DN) (Yaguchi et al., 2004; Pirici et al., 2006; Davidson et al., 2007; Armstrong et al., 2010) (Figure 1).

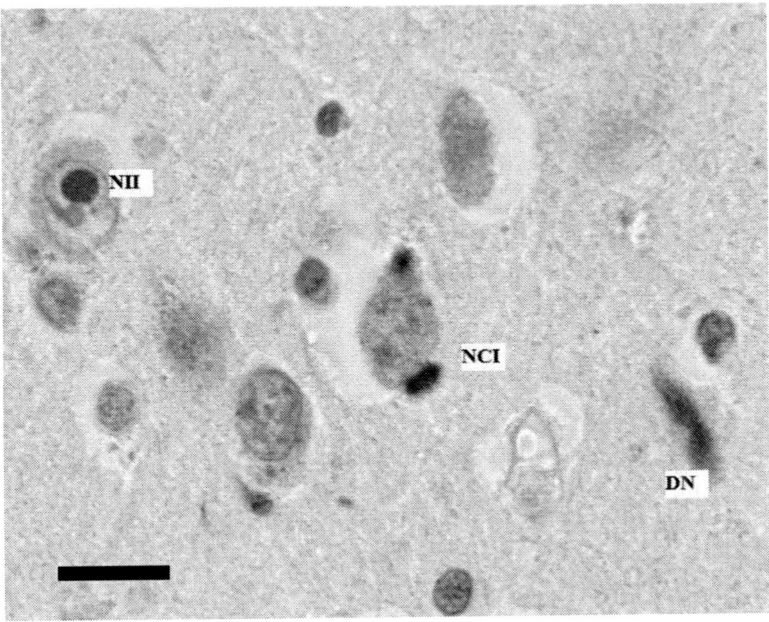

Figure 1. Frontotemporal lobar degeneration with TDP-43 proteinopathy (FTLD-TDP): TDP-43 immunoreactive lesions in the frontal cortex including neuronal cytoplasmic inclusion (NCI), neuronal intranuclear inclusions (NII), and a dystrophic neurite (DN) (TDP-43 immunohistochemistry, bar = 50μm).

FTLD-TDP includes both familial and sporadic cases. Various genetic defects have been identified in the familial cases, the majority being caused by mutation of the *progranulin* (*GRN*) gene (Baker et al., 2006; Cruts et al., 2006; Mukerjee et al., 2006; Behrens et al., 2007; Rademakers and Hutton, 2007). A less prevalent disorder, FTLD with *valosin-containing protein* (VCP) gene mutation (Forman et al., 2006) has TDP-43 immunoreactive inclusions, and familial cases have also been shown to be caused by *ubiquitin associated protein 1 (UBAP1)* (Rollinson et al., 2009), and the *chromosome 9 open reading frame 72 (C9ORF72)* gene (Luty et al., 2008; Renton et al., 2011). In addition, there are familial cases of FTLD not currently associated with any specific gene defect and which do not exhibit a strong autosomal dominant pattern of inheritance (Armstrong et al., 2010).

Cellular inclusions characterized by various molecular pathologies are a common feature of neurodegenerative disease (Goedert et al., 2001, Armstrong and Cairns, 2012). Most disorders are associated with abnormal aggregations of tau (tauopathies), or α-synuclein (synucleinopathies). In the cerebral cortex and hippocampus of these disorders, the various 'signature' pathological inclusions exhibit a distinct spatial pattern, viz., they occur in clusters which exhibit a regular periodicity parallel to the pia mater or alveus respectively (Armstrong et al., 1997; 1998; 1999; 2004; 2011; Armstrong and Cairns, 2012). This spatial pattern suggests that the inclusions may develop in relation to features of the architecture of the cortex and especially to its primary modular structure (Armstrong et al., 2001).

Both phosphorylation-dependent (pTDP-43) and phosphorylation-independent anti-TDP-43 antibodies have been used to study the pathology of FTLD-TDP (Neumann et al., 2009; Hasegawa et al., 2008; Olive et al., 2009; Schwab et al., 2009). The advantage of pTDP-43 antibodies is that they do not immunolabel normal physiological TDP-43 (Neumann et al., 2009; Schwab et al., 2009) thus enabling the TDP-43-immunoreactive inclusions, especially NII, to be more clearly visualized and quantified. In a previous study, the spatial patterns of the NCI were studied using a phosphorylation-independent anti-TDP-43 antibody (Armstrong et al., 2010). Hence, the objective of the present study was to determine in a larger group of cases using a pTDP-43 antibody: (1) whether the phosphorylated inclusions exhibit a similar spatial pattern to that revealed by anti-TDP-43 and (2) whether there are differences in spatial patterns of inclusions in familial and sporadic FTLD-TDP.

MATERIALS AND METHODS

Cases

Thirty-two cases of clinically and neuropathologically verified FTLD-TDP (16 male, 16 female) (see Table 1) were obtained from the Knight Alzheimer's Disease Research Center, Department of Neurology, Washington University School of Medicine, St. Louis, MO, USA. All cases exhibited FTLD with neuronal loss, varying degrees of microvacuolation of the superficial cortical laminae, and a reactive astrocytosis consistent with proposed diagnostic criteria for FTLD-TDP (Cairns et al., 2007b). Of the 32 cases, 20 were familial (at least one or more first degree relatives affected) and of these, 11 cases were caused by *GRN* mutations (Baker et al., 2006; Cruts et al., 2006; Muckerjee et al., 2006; Behrens et al., 2007). The majority (N = 8) of the *GRN* cases come from a single hereditary dysphasic disinhibition dementia (HDDD) family (HDDD2) (Mukherjee et al., 2006) and the remainder (N = 3) from the HDDD1 family (Behrens et al., 2007). Of the remaining familial cases (N = 9), one had a *VCP* gene mutation and one case was associated with the *UBAP1* gene (Rollinson et al., 2009). No genetic defects have been identified to date in the remaining familial cases. Sporadic cases had no evidence of familial involvement and none of the gene mutations known to cause FTLD-TDP were present.

Histological Methods

After death, the consent of the next-of-kin was obtained for brain removal, following local Ethical Committee procedures and the 1995 Declaration of Helsinki (as modified in Edinburgh, 2000). Tissue blocks were taken from the frontal lobe at the level of the genu of the corpus callosum to study the middle frontal gyrus (MFG) (B32) and the temporal lobe at the level of the lateral geniculate body to study the inferior temporal gyrus (ITG) (B20), parahippocampal gyrus (PHG) (B28), CA1/2 sectors of the hippocampus, and dentate gyrus (DG). Tissue was fixed in 10% phosphate buffered formal-saline and embedded in paraffin wax. Immunohistochemistry (IHC) was performed on 4 to 10μm sections with a mouse monoclonal antibody that specifically recognizes phosphorylated pTDP-43 (dilution 1:40,000; pS409/410-1, Clone

11-9, Cosmo Bio USA, Inc., Carlsbad, CA, USA). Sections were also stained with haematoxylin.

Table 1. Summary of demographic features, gross brain weight, and Braak tangle (NFT) score of the 32 cases of frontotemporal lobar degeneration with TDP-43 proteinopathy (FTLD-TDP). (Abbreviations: M/F = male/female, Dur = Disease duration, Death = Age at death, Fm/S = Familial/Sporadic, - indicates data not available, *GRN* = cases caused by *progranulin* gene mutations, *VCP* = case caused by *valosin-containing protein*, *UBAP1* = case caused by *ubiquitin associated protein 1*)

Case	M/F	Onset	Dur	Death	BW	Fm/S	Braak	Subtype
1.	M	57	8	65	960	S	0	2
2.	F	72	12	84	900	S	0	2
3.	F	61	16	77	950	S	1	3
4.	F	68	6	74	975	*GRN*	0	4
5.	M	52	13	65	1300	*GRN*	4	1
6.	M	66	16	82	-	S	-	3
7.	F	69	15	84	970	*GRN*	4	3
8.	M	60	6	66	-	*GRN*	4	1
9.	F	65	12	77	810	*GRN*	4	1
10.	M	52	15	67	960	*GRN*	0	4
11.	M	74	6	80	1270	Fm	4	4
12.	F	-	-	67	990	S	1	2
13.	F	59	9	68	650	Fm	0	2
14.	M	74	1	75	1360	S	1	1
15.	M	60	11	71	1450	Fm	-	4
16.	M	43	7	50	1060	Fm	-	3
17.	M	55	11	66	1005	*GRN*	1	2
18.	F	63	3	66	950	S	-	1
19.	F	58	9	67	880	*GRN*	2	2
20.	F	64	19	83	720	Fm	1	1/2
21.	F	69	4	71	1070	S	1	4
22.	M	38	9	47	1185	*VCP*	0	4
23.	F	-	-	73	720	S	-	1
24.	M	50	18	68	1170	S	0	1
25.	M	58	8	66	1080	Fm	2	4
26.	F	65	13	78	960	Fm	0	-
27.	M	57	6	63	1080	*GRN*	1	-
28.	M	51	11	62	880	*UBAP1*	4	2
29.	F	71	13	84	960	S	2	1
30.	F	73	9	82	-	*GRN*	6	2
31.	F	58	8	66	-	Fm	4	
32.	M	71	8	79	1150	*GRN (2)*	3	

Morphometric Methods

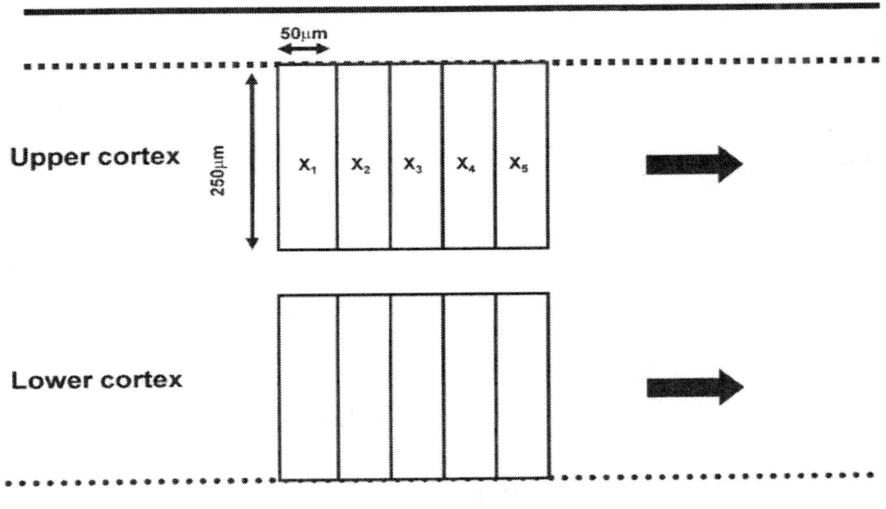

Figure 2. Morphometric method of sampling the cerebral cortex in Frontotemporal lobar degeneration with TDP-43 proteinopathy (FTLD-TDP). Lesions were counted along two strips of tissue parallel to the pia mater (to μm in length) using 200 x 50 μm contiguous sample fields (X_1 to X_5 etc). The sample fields were located both in the upper and lower cortical laminae and extend along the gyrus from sulcus to sulcus; the short edge of the field being aligned either with the surface of the pia mater or with the edge of the white matter.

In each region, NCI, GI, NII, and DN were counted along strips of tissue (1,600 to 3,200 μm in length) located parallel to the pia mater, in 250 x 50 μm sample fields arranged contiguously (Armstrong, 2003) (Fig 2). The sample fields were located both in the upper (approximating laminae II/III) and lower (approximating laminae V/VI) cortex, the short edge of the sample field orientated parallel with the pia mater and aligned with guidelines marked on the slide. In the hippocampus, the inclusions were counted in the cornu ammonis (CA) sectors CA1 and CA2, the short dimension of the contiguous sample field aligned with the edge of the alveus. In addition, NCI have been observed in the dentate gyrus (DG) in FTLD-TDP (Woulfe et al., 2001; Kovari et al., 2004; Mackenzie et al., 2006a) and the sample field was aligned with the upper edge of the granule cell layer. The NCI were round, spicular, or skein-like inclusions (Yaguchi et al., 2004; Davidson et al., 2007), while the GI morphologically resembled the 'coiled bodies' reported in various tauopathies such as corticobasal degeneration (CBD), progressive supranuclear palsy (PSP), and argyrophilic grain disease (AGD). The NII were lenticular or spindle-shaped lesions (Pirici et al., 2006) and the DN characteristically long and contorted (Hatanpaa et al., 2008)

Data Analysis

To determine spatial patterns of the pTDP-43-immunoreactive inclusions, the data were analysed by spatial pattern analysis (Armstrong, 1993a; 1997; 2006). This method uses the variance-mean ratio (V/M) to determine whether the NCI were distributed randomly (V/M = 1), regularly (V/M < 1), or were clustered (V/M > 1) along a strip of tissue parallel to the pia mater. Counts of NCI in adjacent sample fields were added together successively to provide data for increasing field sizes, e.g., 50 x 250 μm, 100 x 250 μm, 200 x 250 μm etc., up to a size limited by the length of the strip sampled. V/M was plotted against field size to determine whether the clusters of inclusions were regularly or randomly distributed and to estimate the mean cluster size parallel to the tissue boundary. A V/M peak indicates the presence of regularly spaced clusters while an increase in V/M to an asymptotic level suggests the presence of randomly distributed clusters. The statistical significance of a peak was tested using the 't' distribution (Armstrong, 1997).

To compare familial and sporadic cases, the 32 cases were divided into three groups: (1) familial FTLD-TDP caused by *progranulin* (*GRN*) mutation, (2) miscellaneous familial cases comprising *valosin-containing protein* (VCP), *ubiquitin associated protein 1 (UBAP1)* cases and cases not associated with any of the currently identified genes, and (3) the remaining sporadic cases. The spatial patterns of the histological features were compared between groups using chi-square (χ^2) contingency table tests. In addition, estimated cluster sizes of inclusions were compared using one-way analysis of variance (ANOVA) followed by Fisher's 'protected least significant difference' (PLSD).

SPATIAL PATTERNS OF pTDP-43-REACTIVE INCLUSIONS

Typical examples of the spatial patterns exhibited by the pTDP-43-immunoreactive inclusions in a single brain region (Case ITG, laminae II/III) are shown in Fig 3. The V/M ratios of the NCI and NII were not significantly different from unity at any field size up to 800 μm suggesting a random distribution of inclusions. By contrast, the GI exhibited a significant V/M peak at a field size of 400 μm suggesting the presence of clusters of inclusions, 400 μm in diameter, regularly distributed parallel to the pia mater. The V/M ratio of the DN increased with field size without reaching a peak suggesting the presence of a large cluster of DN at least 800 μm in diameter.

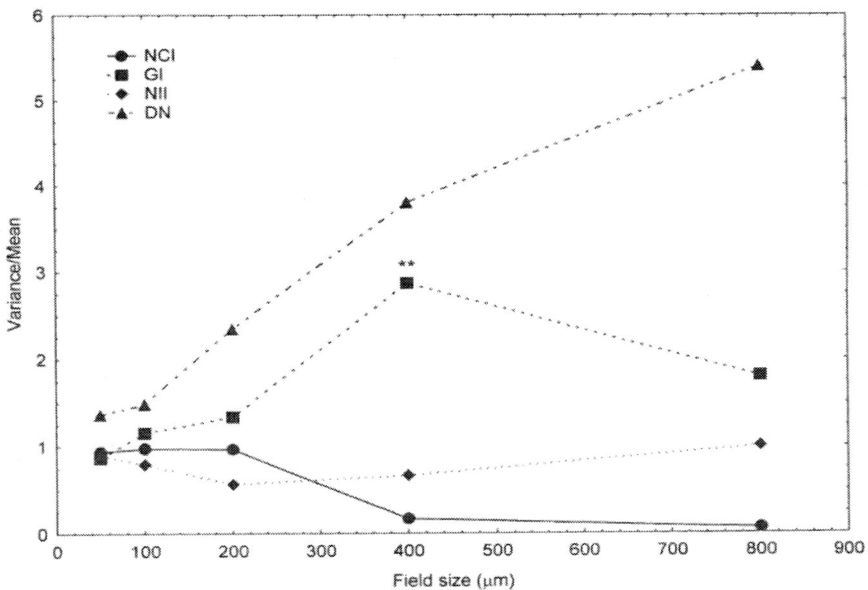

Figure 3. Examples of the spatial patterns exhibited by pTDP-43-immunoreactive inclusions in frontotemporal lobar degeneration with TDP-43 proteinopathy (FTLD-TDP) (** significant V/M peak).

A summary of the spatial patterns exhibited by the pTDP-43-immunoreactive inclusions in all regions in the three groups of cases is shown in Table 2. In many regions, the inclusions developed in clusters, the clusters being regularly distributed parallel to the tissue boundary. This spatial pattern occurred in 60/132 (45%) of analyses of NCI, 33/71 (46%) analyses of GI, 20/51 (39%) analyses of NII, and 44/72 (61%) analyses of DN. Inclusions were also randomly distributed in a proportion of remaining analyses. Clustering on a larger scale, with clusters greater than 800 μm, and a more uniform distribution of inclusions were more infrequent. Within each group of cases, NCI, GI, NII, and DN exhibited a similar range of spatial patterns (GRN mutation cases, $\chi^2 = 9.77$, 9DF, $P > 0.05$; other familial cases $\chi^2 = 8.43$, 9DF, $P > 0.05$; sporadic FTLD-TDP ($\chi^2 = 7.04$, 9DF, $P > 0.05$). In addition, spatial patterns of the NCI ($\chi^2 = 3.11$, 6DF, $P > 0.05$), GI ($\chi^2 = 4.16$, 6DF, $P > 0.05$), ($\chi^2 = 3.73$, 6DF, $P > 0.05$), and DN ($\chi^2 = 3.26$, 6DF, $P > 0.05$) were similar across patient groups. There were no differences between GRN mutation cases originating from the HDDD1 and HDDD2 families and within the miscellaneous familial group between those cases associated with VCP and $UBAP1$ and the remaining familial cases. There were no significant differences in spatial patterns between cortical regions and hippocampus/dentate gyrus or in the cortex, between upper and lower cortical laminae.

Table 2. Summary of spatial patterns exhibited by the histological features (NCI = neuronal cytoplasmic inclusions, GI = glial inclusions, NII = neuronal internuclear inclusions, DN = dystrophic neurites) in progranulin (GRN) mutation cases, other familial cases and sporadic frontotemporal lobar degeneration with TDP-43 proteinopathy (FTLD-TDP) (N = number of regions studied, R = random distribution, Reg = Regular distribution)

Group	Inclusion	N	R	Reg	Reg. Clusters	Large clusters
GRN mutation	NCI	49	16	5	23	5
	GI	27	10	1	15	1
	NII	21	6	2	10	3
	DN	27	4	3	15	5
Remaining familial cases	NCI	36	9	4	16	7
	GI	19	6	3	8	2
	NII	7	2	0	5	0
	DN	19	4	0	13	2
Sporadic cases	NCI	47	11	8	21	7
	GI	25	11	3	10	1
	NII	11	4	0	5	2
	DN	26	5	2	16	3

A comparison of cluster sizes of pTDP-43-immunoreactive inclusions in cerebral cortical regions in familial and sporadic cases is shown in Fig 4. There were no significant differences in mean cluster size of the NCI (F = 1.06, P > 0.05), GI (F = 0.51, P > 0.05), NII (F = 0.97, P > 0.05), or DN (F = 0.48, P > 0.05) between the three patient groups.

The data suggest that in many of the brain regions investigated, the pTDP-43-immunoreactive inclusions occurred in clusters that were regularly distributed parallel to the tissue boundary. This spatial pattern was similar to that reported in FTLD-TDP in which a phosphorylation-independent anti-TDP-43 antibody was used (Armstrong and Cairns, 2010; 2011). In addition, spatial patterns were similar to those characterized by other types of molecular pathology such as tau (Armstrong et al., 1998; 1999), α-synuclein (Armstrong et al., 1997; 2004) and 'fused in sarcoma' (FUS) immunoreactivity (Armstrong et al., 2011). Hence, both phosphorylation-independent and dependent-anti-TDP-43 antibody could be used to investigate the spatial patterns of inclusions in FTLD-TDP. In addition, spatial patterns of cellular inclusions appear to be independent of molecular subtype of disease (Armstrong and Cairns, 2012).

Chi-square (χ^2) contingency table analysis: (1) Between inclusions within each group: GRN cases $\chi^2 = 9.77$ (9DF, P > 0.05), Other familial cases $\chi^2 = 8.43$ (9DF, P > 0.05), sporadic cases $\chi^2 = 7.04$ (9DF, P > 0.05), (2) Between patient groups for each inclusion: NCI $\chi^2 = 3.11$ (6DF, P > 0.05), GI $\chi^2 = 4.16$ (6DF, P > 0.05), NII $\chi^2 = 3.73$ (6DF, P > 0.05), DN $\chi^2 = 3.26$ (6DF, P > 0.05)

A major feature of the architecture of the cerebral cortex is the replicated local neural circuit (represented by 'columns' or 'modules') (Hiorns et al., 1991). The diameter of individual cortical modules varies between 500-1000μm depending on region and there are specific connections maintained between ordered sets of columns (Hiorns et al., 1991). In disorders characterized by tau, α-synuclein, and FUS immunoreactivity, the spatial patterns of the various inclusions within the cerebral cortex and hippocampus suggested that the inclusions could be related to this modular structure, most specifically, the cells of origin of specific cortico-cortical and cortico-hippocampal projections (Delacoste and White, 1993; Hiorns et al., 1991; Delatour et al., 2004; Armstrong and Cairns, 2012). First, the cells of origin of the cortico-cortical projections are clustered and occur in bands that are more or less regularly distributed along the cortical strip. Second, individual bands of cells, approximately 500 – 800 μm in width, depending on cortical region, traverse the cortical laminae in columns (Hiorns et al., 1991). In 44 - 66% of cortical regions studied, the clusters of pTDP-43 immunoreactive inclusions were regularly distributed parallel to the pia mater consistent with their development in relation to these connections. Nevertheless, in only a small number of cortical areas did the estimated width of the NCI clusters approximate to the dimension of the cells of origin of the cortico-cortical projections. In the majority of cortical areas, the NCI developed in clusters, 50 - 200 μm in diameter, smaller than the size of these columns and a size similar to the FUS-immunoreactive NCI in neuronal intermediate filament inclusion disease (NIFID) (Armstrong et al., 2011). Hence, pTDP-43 immunoreactive inclusions may affect only a subset of cells within a modular column. In addition, in some regions, clusters larger than 800 μm in diameter were present suggesting that smaller clusters of inclusions, especially of the GI and NII, could develop into larger aggregations as the disease develops (Armstrong, 1993b). Hence, formation of clusters of TDP-43-immunoreactive inclusions may be a dynamic process changing as the disease develops.

The spatial pattern of the pTDP-43-immunoreative inclusions was similar in *GRN* mutation cases, miscellaneous familial cases, and in sporadic FTLD-TDP. Several different frame-shift and premature termination mutations have been identified in FTLD-TDP with *GRN* mutation (Beck et al., 2008). Abnormal protein products may accumulate within the endoplasmic reticulum of the cell due to inefficient secretion or mutant RNA may have a lower expression within the cell at least in some mutants (Mukherjee et al., 2006). TDP-43 is a nuclear protein but in FTLD-TDP, TDP-43 is redistributed from the nucleus to the cytoplasm, is ubiquinated, hyperphosphorylated, and then cleaved to generate C-terminal fragments (Neumann et al., 2007). These fragments accumulate to form the NCI and NII and eventually cause cell death. Analogous mechanisms may be involved in other types of familial case whereas the molecular mechanism responsible for pTDP-43 accumulation in sporadic cases has not been established. Hence, there is no unique pattern of spatial distribution of pTDP-43 inclusions in familial FTLD-TDP. Pathological changes initiated by *GRN* mutation, by other genetic factors in the remaining familial cases, and by other causes in sporadic FTLD-TDP appear to follow a parallel course resulting in very similar patterns of cortical degeneration in frontal and temporal lobes.

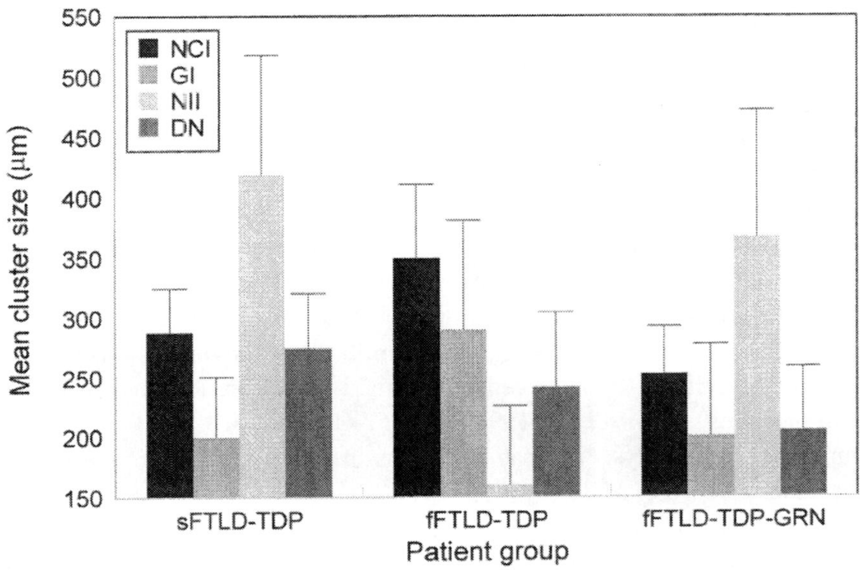

Figure 4. A comparison of cluster sizes of neuronal cytoplasmic inclusions (NCI), glial inclusions (GI), neuronal internuclear inclusions (NII), and dystrophic neuritis (DN) in progranulin (GRN) mutation cases, other familial cases and sporadic frontotemporal lobar degeneration with TDP-43 proteinopathy (FTLD-TDP). Analysis of Variance (ANOVA): NCI (F = 1.06, P > 0.05), GI (F = 0.51, P > 0.05), NII (F = 0.97, P > 0.05), DN (F = 0.48, P > 0.05).

In conclusion, the pTDP-immunoreactive inclusions in FTLD-TDP are largely distributed in regular clusters, a spatial pattern similar to that reported for inclusions characterized by tau, α-synuclein, and FUS immunoreactivity. The spatial patterns are also similar to those reported in FTLD-TDP cases using a phosphorylation-independent antibody. This spatial pattern was also consistently present in familial and sporadic cases of FTLD-TDP suggesting similar patterns of cortical degeneration.

ACKNOWLEDGMENTS

I thank clinical, genetic, pathology, and technical staff of the Departments of Neurology and Pathology and Immunology, Washington University School of Medicine, St Louis, MO, USA for making available cases for this study.

REFERENCES

Amandor-Ortiz C, Lin WL, Ahmed Z, Personett D, Davies P, Duara R, Graff-Radford NR, Hutton ML, Dickson DW. TDP-43 immunoreactivity in hippocampal sclerosis and Alzheimer's disease. *Ann Neurol* 2007, 61, 435-445.

Armstrong RA. The usefulness of spatial pattern analysis in understanding the pathogenesis of neurodegenerative disorders, with particular reference to plaque formation in Alzheimer's disease. *Neurodegen* 1993a, 2, 73-80.

Armstrong RA. Is the clustering of neurofibrillary tangles in Alzheimer's patients related to the cells of origin of specific cortico-cortical projections? *Neurosci Lett* 1993b, 160, 57-60.

Armstrong RA. Analysis of spatial patterns in histological sections of brain tissue. *J Neurosci Meth* 1997, 73, 141-147.

Armstrong RA. Quantifying the pathology of neurodegenerative disorders: quantitative measurements, sampling strategies and data analysis. *Histopathol* 2003, 42, 521-529.

Armstrong RA. Methods of studying the planar distribution of objects in histological sections of brain tissue. *J Microsc* (Oxf) 2006, 221, 153-158.

Armstrong RA, Cairns NJ, Lantos PL. Dementia with Lewy bodies: Clustering of Lewy bodies in human patients. *Neurosci Lett* 1997, 224, 41-44.

Armstrong RA, Cairns NJ, Lantos PL. Clustering of Pick bodies in Pick's disease. *Neurosci Lett* 1998, 242, 81-84.

Armstrong RA, Cairns NJ, Lantos PL. Clustering of cerebral cortical lesions in patients with corticobasal degeneration. *Neurosci Lett* 1999, 268, 5-8.

Armstrong RA, Cairns NJ, Lantos PL. What does the study of spatial patterns tell us about the pathogenesis of neurodegenerative disorders? *Neuropathology* 2001, 21, 1-12

Armstrong RA, Lantos PL, Cairns NJ. Spatial patterns of α-synuclein positive glial cytoplasmic inclusions in multiple system atrophy. *Move Disorders* 2004, 19, 109-112.

Armstrong RA, Ellis W, Hamilton RL, Mackenzie IRA, Hedreen J, Gearing M, Montine T, Vonsattel J-P, Head E, Lieberman AP, Cairns NJ. Neuropathological heterogeneity in frontotemporal lobar degeneration with TDP-43 proteinopathy: a quantitative study of 94 cases using principal components analysis. *J Neural Transm* 2010, 117, 227-239.

Armstrong RA, Cairns NJ. A morphometric study of the spatial patterns of the TDP-43-immunorecative neuronal inclusions in frontotemporal lobar degeneration with *progranulin* (*GRN*) mutation. *Histol Histopathol* 2011, 26, 185-190.

Armstrong RA, Gearing M, Bigio EH, Cruz-Sanchez FF, Duyckaerts C, Mackenzie IRA, Perry RH, Skullerud K, Yokoo H, Cairns NJ. Spatial patterns of FUS-immunoreactive neuronal cytoplasmic inclusions (NCI) in neuronal intermediate filament inclusion disease. *J Neural Transm* 2011, 118, 1651-1657.

Armstrong RA, Cairns NJ. Different molecular pathologies result in similar spatial patterns of cellular inclusions in neurodegenerative disease: a comparative study of eight disorders. *J Neural Transm* 2012, 119, 1551-1560.

Baker M, Mackenzie IR, Pickering-Brown SM, Gass J, Rademakers R, Lindholm C, Snowden J, Adamson J, Sadovnick AD, Rollinson S, Cannon A, Dwosh E, Neary D, Melquist S, Richardson A, Dickson D, Berger Z, Eriksen J, Robinson T, Zehr C, Dickey CA, Crook R, McGowan E, Mann D, Boeve B, Feldman H, Hutton M. Mutations in progranulin cause tau-negative frontotemporal dementia linked to chromosome 17. *Nature* 2006, 442, 916-919.

Beck J, Rohrer JD, Campbell T, Isaacs A, Morrison KE, Goodall EF, Warrington EK, Stevens J, Revesz T, Hoton J, Al-Sarraj S, King A, Scabill R, Warren JD, Rossor MN, Collinge J, Mead S. A distinct clinical, neuropsychological and radiological phenotype is associated with progranulin gene mutation in a large UK series. Brain 2008, 131, 706-720.

Behrens MI, Mukherjee O, Tu PH, Liscic RM, Grinberg LT, Carter D, Paulsmeyer K, Taylor-Reinwald L, Gitcho M, Norton JB, Chakraverty S, Goate AM, Morris JC, Cairns NJ. Neuropathologic heterogeneity in HDDD1: a familial frontotemporal lobar degeneration with ubiquitin-positive inclusions and progranulin mutation. *Alz Dis Assoc Disord* 2007, 21, 1-7.

Brandmeir NJ, Geser F, Kwong LK, Zimmerman E, Qian J, Lee VMY, Trojanowski JQ. Severe subcortical TDP-43 pathology in sporadic frontotemporal lobar degeneration with motor neuron disease. *Acta Neuropathol* 2008, 115, 123-131.

Cairns NJ, Neumann M, Bigio EH, Holm IE, Troost D, Hatanpaa KJ, Foong C, White CL III, Schneider JA, Kretzschmar HA, Carter D, Taylor-Reinwald L, Paulsmeyer K, Strider J, Gitcho M, Goate AM, Morris JC, Mishra M, Kwong LK, Steiber A, Xu Y, Forman MS, Trojanowski JQ, Lee VMY, Mackenzie IRA. TDP-43 familial and sporadic frontotemporal lobar degeneration with ubiquitin inclusions. *Am J Pathol* 2007a, 171, 227-240.

Cairns NJ, Bigio EH, Mackenzie IRA, Neumann M, Lee VMY, Hatanpaa KJ, White CL, Schneider JA, Grinberg LT, Halliday G, Duyckaerts C, Lowe JS, Holm IE, Tolnay M, Okamoto K, Yokoo H, Murayama S, Woulfe J, Munoz DG, Dickson DW, Ince PG, Trojanowski JQ, Mann DMA. Neuropathologic diagnostic and nosological criteria for frontotemporal lobar degeneration: consensus of the Consortium for Frontotemporal Lobar Degeneration. *Acta Neuropathol* 2007b, 114, 5-22.

Cruts M, Gijselink I, van der ZJ, Engelborgs S, Wils H, Pirici D, Radamakers R, Vandenberghe R, Dermaut B, Martin JJ, van Duijn C, Peeters K, Sciot R, Santens P, De pooter T, Mattheijssens M, van den BM, Cuijt I, Vennekens K, De Deyn PP, Kumar-Singh S, Van Broeckhoven C. Null mutations in progranulin cause ubiquitin-positive frontotemporal dementia linked to chromosome 17q21. *Nature* 2006, 442, 920-924.

Davidson Y, Kelley T, Mackenzie IRA, Pickering Brown S, Du Plessis D, Neary D, Snowden JS, Mann DMA. Ubiquinated pathological lesions in frontotemporal lobar degeneration contain TAR DNA-binding protein, TDP-43. *Acta Neuropathol* 2007, 113, 521-533.

De Lacoste M, White CL III (1993) The role of cortical connectivity in Alzheimer's disease pathogenesis: a review and model system. *Neurobiol Aging* 14, 1-16.

Delatour B, Blanchard V, Pradier L, Duyckaerts C. Alzheimer pathology disorganizes cortico-cortical circuitry: direct evidence from a transgenic animal model. *Neurobiol Dis* 2004, 16, 41-47.

Forman MS, Mackenzie IR, Cairns NJ, Swanson E, Boyer PJ, Drachman DA, Jhaveri BS, Karlawish JH, Pestrvik A, Smith TN, Tu PH, Watts GDJ, Markesbery WR, Smith CD, Kimonis VE. Novel ubiquitin neuropathology in frontotemporal dementia with valosin-containing protein gene mutations. *J Neuropathol Exp Neurol* 2006, 65, 571-581.

Goedert M, Spillantini MG, Serpell LC, Berriman J, Smith MJ, Jakes R, Crowther RA. From genetics to pathology: tau and alpha-synuclein assemblies in neurodegenerative diseases. *Phil Trans Roy Soc* (Lond) B. Biol Sci 2001, 356, 213-227.

Hasegawa M, Arai T, Akiyama H, Noriaka T, Mori H, Hashimoto T, Yamazaki M, Oyanagi K. TDP-43 is deposited in the Guam parkinsonism-dementia complex brain. *Brain* 2007, 130, 1386-1394.

Hatanpaa KJ, Bigio EH, Cairns NJ, Womack KB, Weintraub S, Morris JC, Foong C, Xiao GH, Hladik C, Mantanona TY, White CL. TAR DNA-binding protein 43 immunohistochemistry reveals extensive neuritic pathology in FTLD-U: A Midwest-Southwest Consortium for FTLD-U study. *J Neuropathol Exp Neurol* 2008, 67, 271-279.

Higashi S, Iseki E, Yamamoto R, Minegishi M, Hino H, Fujisawa K, Togo T, Katsuse O, Uchikado H, Furiukawa Y, Kosaka K, Arai H. Concurrence of TDP-43, tau and □-synuclein pathology in brains of Alzheimer's disease and dementia with Lewy bodies. *Brain Res* 2007, 1184, 284-294.

Hiorns RW, Neal JW, Pearson RCA, Powell TPS. Clustering of ipsilateral cortico-cortical projection neurons to area 7 in the rhesus monkey. *Proc Roy Soc Lond* 1991, 246, 1-9.

Kovari E, Gold G, Giannakopoulos P, Bouras C. Cortical ubiquitin positive inclusions in frontotemporal dementia without motor neuron disease: a quantitative immunocytochemical study. *Acta Neuropathol* 2004, 108, 207-212.

Luty AA, Kwok JBJ, Thompson EM, Blumsbergs P, Brooks WS, Loy CT, Dobson-Stone C, Panegyres PK, Hecker J, Nicholson GA, Halliday GM, Schofield PR. Pedigree with frontotemporal lobar degeneration-motor neuron disease and Tar DNA binding protein-43 positive neuropathology: genetic linkage to chromosome 9. *BMC Neurology* 2008, 8, 32.

Mackenzie IRA, Baker M, Pickering-Brown S, Hsinng GYR, Lindholm C, Dwosh E, Cannon A, Rademakers R, Hutton M, Feldman HH. The neuropathology of frontotemporal lobar degeneration caused by mutations in the progranulin gene. *Brain* 2006, 129, 3081-3090.

Mukherjee O, Pastor P, Cairns NJ, Chakraaverty S, Kauwe JSK, Shears S, Behrens MI, Budde J, Hinrichs AL, Norton J, Levitch D, Taylor-Reinwald L, Gitcho M, Tu PH, Grinberg LT, Liscic RM, Armendariz J, Morris JC, Goate AM. HDDD2 is a familial frontotemporal lobar degeneration with ubiquitin-positive tau-negative inclusions caused by a missense mutation in the signal peptide of progranulin. *Ann Neurol* 2006, 60,314-322.

Neumann M, Igaz LM, Kwong LK, Nakashima-Yasuda H, Kolb SJ, Dreyfuss G, Kretzschmar HA, Trojanowski JQ, Lee VMY. Absence of heterogeneous nuclear riboproteins and survival neuron protein (TDP-43) positive inclusions in frontotemporal lobar degeneration. *Acta Neuropathol* 2007, 113, 543-548.

Neumann M, Kwong LK, Lee EB, Kremmer E, Flately A, Xu Y, Forman MS, Troost D, Kretzschmar HA, Trojanoswki JQ, Lee VMY. Phosphorylation of S409/410 of TDP-43 is a consistent feature in all sporadic and familial forms of TDP-43 proteinopathies. *Acta Neuropathol* 2009, 117, 137-149.

Olive M, Janve A, Moreno D, Gamez J, Torrejon-Escritano B, Ferrer I. TAR DNA-binding protein 43 accumulation in protein aggregate myopathies. *J Neuropathol Exp Neurol* 2009, 68, 262-273.

Pirici D, Vandenberghe R, Rademakers R, Dermant B, Cruts M, Vennekens K, Cuijt I, Lubke U, Centerick C, Martin JJ, Van Broeckhoven C, Kumar-Singh S. Characterization of ubiquinated intraneuronal inclusions in a novel Belgian frontotemporal lobar degeneration family. *J Neuropath Exp Neurol* 2006, 65, 289-301.

Rademakers R, Hutton M. The genetics of frontotemporal lobar degeneration. *Curr Neurol Neurosci Rep* 2007, 7, 434-442.

Renton AE, Majounie E, Waite A, Simón-Sánchez J, Rollinson S, Gibbs JR, Schymick JC, Laaksovirta H, van Swieten JC, Myllykangas L, Kalimo H, Paetou A, Abramzon Y, Remes AM, Kaganovitch A, Scholz SW, Duckworth J, Ding J, Harmer DW, Hernandez DG, Johnson JO, Mok K, Ryten M, Trabzuni D, Guerreiro RJ, Orrell RW, Neal J, Murray A, Pearson J, Jansen IE, Sondervan D, Seelaar H, Blake D, Young K, Halliwell N, Callister JB, Toulson G, Richardson A, Gerhard A, Snowden J, Mann D, Neary D, Nalls MA, Peuralinna T, Jansson L, Isoviita VM, Kalvorinne AL, Hölttä-Vuori M, Ikonen E, Sulkava R, Benatar M, Wuu J, chio A, Restagno G, Borghero G, Sabatelli M, The ITALSGEN Consortium, Heckerman D, Rogaeva E, Zinman L, Rothstein JD, Sendtner M, Drepper C, Eichler EE, Alkan C, Abdullaev Z, Pack SD, Dutra A, Pak E, Hardy J, Singleton A, Williams NM, Heutink P, Pickering-Brown S, Morris HR, Tienari PJ, Traynor BJ (2011). A hexanucleotide repeat expansion in *C9ORF72* is the cause of chromosome 9p21-linked ALS-FTD. *Neuron* 2011, 72, 257-268.

Rollinson S, Rizzu P, Sikkink S, Baker M, Halliwell N, Snowden J, Traynor BJ, Ruano D, cairns N, Rohrer JD, Mead S, Collinge J, Rossor M, Akay E, Gueireiro R, Rademakers R, Morrison KE, Pastor P, Alonso E, Martinez-Lage P, Graff-Radford N, Neary D, Henlink P, Mann DMA, Van Swieten J, Pickering-Brown SM. Ubiquitin associated protein 1 is a risk factor for frontotemporal lobar degeneration. *Neurobiol Aging* 2009, 30, 656-665.

Schwab C, Arai T, Hasegawa M, Akiyama H, Yu S, McGeer PL. TDP-43 pathology in familial British dementia. *Acta Neuropathol* 2009, 118, 303-311.

Uryu K, Nakashima-Yasuda H, Forman MS, Kwong LK, Clark CM, Grossman M, Miller BL, Kretzschmar HA, Lee VM, Trojanowski JQ, Neumann M. Concomitant TDP-43 pathology is present in Alzheimer's disease and corticobasal degeneration but not in other tauopathies. *J Neuropathol Exp Neurol* 2008, 67, 555-564.

Woulfe J, Kertesz A, Munoz DG. Frontotemporal dementia with ubiquitinated cytoplasmic and intranuclear inclusions. *Acta Neuropathol* 2001, 102,94-102.

Wang HY, Wang IF, Bose J, Shen CKJ. Structural diversity and functional implications of the eukaryotic TDP gene family. *Genomics* 2004, 83, 130-139.

Yaguchi M, Fujita Y, Amari M, Takatama M, Al-Sarraj S, Leigh PN, Okamoto K. Morphological differences of intraneural ubiquitin positive inclusions in the dentate gyrus and parahippocampal gyrus of motor neuron disease with dementia. *Neuropathol* 2004, 24, 296-301.

In: Horizons in Neuroscience Research. Vol. 25   ISBN: 978-1-63485-286-9
Editors: A. Costa and E. Villalba   © 2016 Nova Science Publishers, Inc.

Chapter 6

NEUROPATHOLOGICAL ASPECTS OF ACUTE DISSEMINATED ENCEPHALOMYELITIS

Shinji Ohara, MD

Department of Neurology, Matsumoto Medical Center, Matsumoto, Japan

ABSTRACT

Acute disseminated encephalomyelitis (ADEM) has been generally considered as an immunologically-mediated inflammatory demyelinating disease of the CNS. Unlike Multiple Sclerosis (MS), which is characterized by clinical relapses and progression over years, ADEM is usually monophasic and relatively with a good prognosis. Although, ADEM and MS have been long considered as separate disease entities, clinical differentiation of ADEM from the first attack of MS is often difficult because of overlapping clinical features. Pathologically, perivenous demyelination and discrete confluent demyelination have been generally regarded as the hallmark of ADEM and MS, respectively. However, hybrid cases showing cardinal pathological features of both ADEM and acute stages of MS do exist, suggesting that ADEM may share some common underlying pathologic mechanisms with certain stages or subgroups of MS. Some patients with clinically or pathologically diagnosed ADEM may later develop MS, although others apparently do not. Factors that may be responsible for such differences

---

* Correspondence to: Dr.Shinji Ohara, Department of Neurology, Matsumoto Medical Center, Chushin-Matsumoto Hospital, 811 Kotobuki, Matsumoto, 399-0021 Japan. Tel +81-263-583121 Fax+81-263-863190. Email; oharas@hosp.go.jp.

are currently unknown, although the presence of certain acquired mechanisms to suppress later disease development may be a possibility. Moreover, cases of ADEM with concomitant peripheral nerve involvement have been documented clinically and pathologically, which may suggest common features underlying immune etiologies involving both CNS and PNS.

Keywords: acute disseminated encephalomyelitis, multiple sclerosis, perivenous demyelination, confluent demyelination, microglia, CD8+ T lymphocyte, Guillain-Barre syndrome

INTRODUCTION

ADEM and MS have been considered clinically and pathologically distinct phenotype of inflammatory demyelinating disorder of the CNS. Typically, ADEM is a monophasic illness, while the most common form of in MS is characterized by a relapsing and remitting course. Pathologically, ADEM is characterized by numerous foci of demyelination scattered in the CNS, invariably surrounding small and medium-sized veins. Perivenular inflammatory infiltrates of lymphocytes and mononuclear cells are another distinctive feature, and multifocal meningeal infiltration is invariably present [1]. Occurrence of subpial lesions in ADEM has been previously described in the literature [2], and more recently described [3]. However, despite several clinical and/or radiological criteria proposed to differentiate ADEM from MS, none have been proven to unequivocally separate them. With many similarities in clinical presentation, in MRI findings and in putative pathogenesis, it is speculated that ADEM may not be a distinctive disease but a part of the MS spectrum [4].

We have recently experienced such a hybrid case clinically diagnosed as ADEM whose brain biopsy revealed pathological features indistinguishable from active lesions of MS in addition to characteristic foci of perivenular inflammation and demyelination of ADEM [5]. The patient made an excellent recovery with steroid treatment and there has been no recurrence in 8.5 years of follow-up (at the time of this writing).

Relatively few cases of ADEM with immunohistochemical analysis have been available so far, and it remains unsettled if ADEM is immunopathologically distinct from MS or may share common underlying pathologic mechanisms with MS. In this review, we will introduce our recent clinical case [5] with some additional pathological observations and review the literature from a neuropathological standpoint.

PRESENTATION OF A CASE

Detailed clinical information of this patient has been reported previously [5]. In short, a 51-year female presented with progressive aphasia and right-sided hemiparesis after a month history of new onset increasing headache. There was no apparent antecedent flu-like illness or history of vaccination. Cerebrospinal fluid (CSF) examination revealed slight pleocytosis and increase in protein and myelin basic protein (MBP). There were no oligoclonal bands. Brain MRI revealed multifocal subcortical lesions (many of them more than 2 cm in diameter) mainly involving subcortical white matter of the left temporal and parietal lobes without enhancement.

Because of the rapid progression of her symptoms with the new onset of focal seizures, a brain biopsy was performed. She was then started on intravenous methyl prednisolone followed by oral prednisolone taper. The patient responded remarkably both clinically and radiologically, and was discharged two month later. She was able to resume her former daily social activities. The last MRI examination, 6 years after the onset of initial symptoms, revealed mild atrophic changes with gliosis involving temporal and parietal lobes on the right side, with no evidence of any new lesions. At the time of this writing, 8.5 years later, there has been no clinical recurrence.

The brain biopsy specimen revealed multiple demyelinating lesions in the he junctional and subcortical white matter. Two patterns could be recognized; the first consisting of a restricted rim of perivenous demyelination with perivascular cuffing, and the second plaque-like confluent demyelination. The leptomeninges revealed an inflammatory cell infiltrates, consisting of B and T lymphocytes. Areas of subpial and intracortical demyelination were also identified by sensitive immunohistochemistry with antibodies to MBP and CNPase. In the confluent demyelinating lesions, increase in the number of macrophages/microglia and clusters of reactive oligodendrocytes were recognized. There was no deposition of IgG lambda or kappa light chains in the demyelinating lesions or in the perivascular cuff [5].

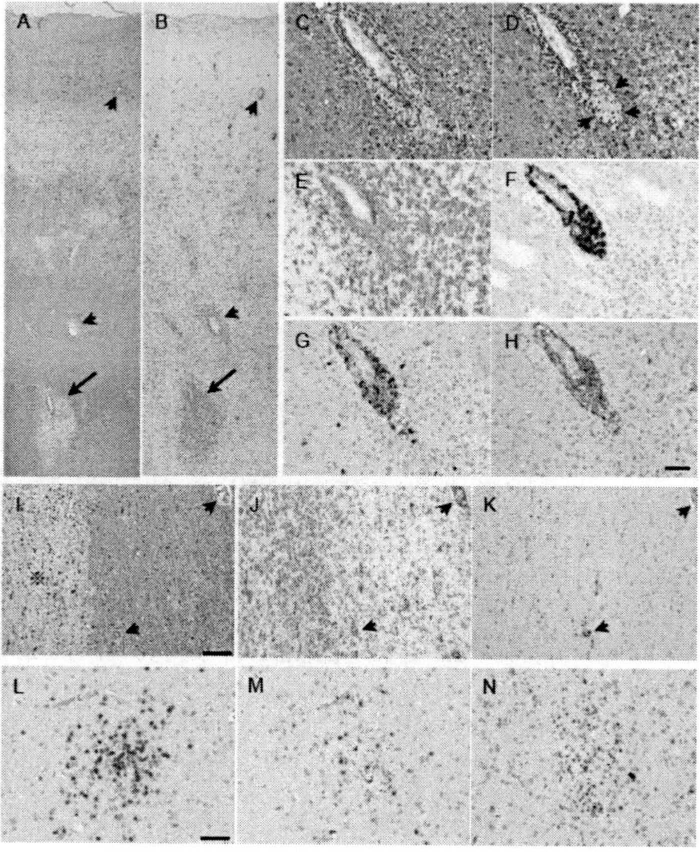

Brain biopsy findings. (A, B) serial sections of the specimen containing cortex and subcortical white matter, stained with antibodies for CNPase (A) and HLA-DR (B). In this low power micrograph, areas of perivenular demyelination (arrow) corresponds to increased density of HLD-DR positive activated microglia. Activated microglia are diffusely seen in the deeper layer of the cortex and adjacent subcortical white matter with perivascular accentuation not associated with decreased CNPase immunostaining (arrowheads). (C-H) serial sections of the area indicated with arrow in (A, B), immunostained with antibody to neurofilament (C), GFAP (D), HLA-DR (E), CD20 (F), CD8 (G), and CD4 (H) respectively. Neurofilaments (C) are well preserved in the area of perivascular demyelination where increase in the number of HLA-DR positive microglia (E) and reactive astrocytes (D) are seen. Arrowheads indicate glia limitans defining perivascular infiltrates. CD20+ B cells (F) and CD4+ T cells (H) tend to be restricted in the perivascular space, while CD8+ T cells (G) are more scattered in the area of perivascular demyelination. (I-K) These micrographs show the area of confluent demyelination indicated with (※) immunostained with antibody for neurofilament (I), HLA-DR (K) and CD8+ (L). The increase in the number of HLA-DR positive activated microglia is most evident in the area of demyelination and lesser in degree in the neighboring white matter. Arrowheads indicate perivascular infiltration of CD8+ and HLA-DR positive cells. (L-N) These micrographs illustrate clustering of T (L, CD8+; M, CD4+) and B lymphocytes (N, CD20+) occasionally seen in the white matter showing ill-defined area of demyelination. In addition to CD8+ and CD4+ cells, CD20+ cells are also found although less in number.

Bar= 50 μ m for (C-H, and L-M), 100 μ m for (I-K).

To further clarify pathological similarities or dissimilarities of this case from MS, further immunohistochemical studies were performed with special attention to microglia and subset of T lymphocytes (CD8+, CD4+), using mouse monoclonal antibodies against HLA-DR (DAKO 1:100), CD4 (Novocastra 1:400), CD8 (Novocastra, 1:200).

Immunostaining for HLA-DR revealed that immunoreactive cells (activated microglia) were most populated in perivascular (Figure B) and also in the confluent foci of demyelination (Figure J). In perivascular cuff, they were often closely related to CD20+ B cells (Figure F), CD8+ T cells (Figure G) and CD4+ T cells (Figure H). Unlike CD8+ T cells which were found infiltrating in the demyelinating lesions (Figure G, K), CD20+ B cells tended to be restricted in the perivascular space limited outward by the glia limitans (Figure D, F). The microglia with ramifying HLA-DR positive cell processes were also increased in number in the neighborhood of demyelinating lesions where there were no apparent features of demyelination or axonal loss (Figure B). Clusters of CD8+ cells could be occasionally seen in the area showing modest demyelinating changes. In one site, they were intermingled with lesser numbers of CD4+ T cells and CD20+ B cells (Figure L, M, N). Features suggestive of oligodendrocyte degeneration such as the appearance of pyknotic nuclei in the periphery of the demyelinating lesions were not apparent. Rather, activated oligodendrocytes with rich CNPase immunoreactive cytoplasm were most abundant in the confluent demyelinating lesions, and to a lesser degree, also found in the areas of perivascular demyelination.

DISCUSSION

Comments on the Present Case

The present case could be regarded clinically as an intermediate form of ADEM and tumefactive MS based on the presence of multiple large subcortical white matter demyelinating lesions and atypical clinical features for MS including focal cortical signs and seizures [6, 7]. Brain biopsy from the MRI identified active lesion showed both patterns of perivenous and confluent patterns of demyelination, which has been regarded as a pathologic gold standard of ADEM and MS respectively [3]. Moreover, the demyelinating lesions of the present case share other important features of active MS lesions including infiltration of T and B lymphocytes and microglial activation in the lesions.

In MS, inflammation is dominated by CD8 positive T cells and the number of CD20 positive B cells is much lower [8-10]. T cells are seen in perivascular cuffs but they also infiltrate into CNS parenchyma. By contrast, B cells tend to accumulate in perivascular space and meninges [11]. Essentially, identical features are recognized in the demyelinating lesions in the present case. In ADEM, relatively few immunohistochemical studies have been performed on the nature of infiltrating cells, describing variability in the appearance and nature of T and B lymphocytes [12]. In one fulminant case of ADEM, stereotaxic biopsy of the frontal white matter showed perivenous inflammation and demyelination with presence of CD3+, CD8+, but not CD20+ B cells [13].

In MS, activated HLA-DR expressing microglia are not only seen in the demyelinating lesions but also in the vicinity of active demyelinating lesions (referred to as preactive lesions) and even in normal appearing white matter (NAWM). They were not associated the T cell infiltrates, axonal alteration, activated astrocytes and blood brain barrier (BBB) disruption [14]. It is considered that the activation of microglia may represent a stage preceding the emergence of inflammatory lesions [15], or it could be nonspecific to MS and is associated with degenerating as well as damaged axons in early MS [16]. In the present study, activated microglia were invariably found in the active demyelinating lesions in the same manner as in MS. They could also be recognized in perivascular cuffs with or without coexisting lymphocytes and in the non-demyelinating white matter near the demyelinating lesions.

It should be noted, unlike ADEM with relatively homogenous histology, the pathology of MS is heterogeneous and may reflect different underlying mechanisms of demyelination. According to the most recent classification scheme by Lucchinetti et al. four histologic patterns could be separated in MS demyelinating lesions: Pattern I lesions show inflammatory lesions made up of T cells and macrophages alone: Pattern II lesions contain immunoglobulin and complement: Pattern III is a distal dying back oligodendrogliopathy: and Pattern IV is characterized by primary oligodendroglial degeneration [17]. The demyelinating pattern of the case presented here is most consistent with Pattern I, in that microglia/macrophage and T cells infiltration was evident in the active demyelinating lesions in the absence of IgG deposition by immunohistochemistry. The degenerative changes of oligodendroglia were not apparent. Rather, there were abundant reactive oligodendrocytes in the demyelinating lesions, which may be compatible with patient's excellent clinical recovery.

DO MS AND ADEM BELONG TO A SPECTRUM OF A SINGLE DISEASE?

To date, there are no clear clinical or radiologic parameters to discriminate ADEM and MS, despite several clinical and/or radiological criteria proposed. In the series of Schwarz [18], 35% of the 40 patients developed clinically definite MS (Poser criteria) over a mean observation of 38 months. de Seze et al. reported that out of 60 patients initially diagnosed as ADEM, 32 % were subsequently diagnosed as definite MS after a mean follow up of 37 months [19]. Thus, the only truly reliable diagnostic test is time [20, 21]. On the other hand, pathological differentiation between MS and ADEM has been regarded as most reliable; MS is characterized by confluent demyelination and ADEM by privenous demyelination, with these two patterns seldom coexisting in a single patient. Hoche et al. reported a 16 year-old patient clinically diagnosed as ADEM with histological features of active multiple sclerosis (pattern I according to Lucchinetti) on brain biopsy, showing confluent demyelination and no findings supportive of ADEM such as a perivenular demyelination [22]. Without tissue examination, this case would have been simply diagnosed as ADEM. Guenther et al. reported an autopsy case of a 19 year-old clinically diagnosed as ADEM who suddenly died of sepsis. Autopsy revealed multiple foci in a confluent demyelination pattern in the absence of areas of perivenular demyelination. The author interpreted that the case could represent either a first demyelinating event in a case of MS with ADEM-like clinical presentation or a case of ADEM with an unusual MS-like pattern of demyelination [23]. Yidiz et al. reported a 35 year-old woman with 14-year history of relapsing-remitting MS who developed a hyperacute form of ADEM (Acute Hemorrhagic Leukoencephalitis AHLE), a diagnosis confirmed by brain biopsy. Interestingly, this patient had started experiencing typical MS relapses after the ALHE episode had resolved. The authors believe that MS and ADEM occurred independently in this patient [24].

However, together with these cases illustrating the clinical overlap of MS and ADEM, the presence of hybrid cases with overlapping pathological features strongly argues that MS and ADEM represent both ends of the spectrum of a single disease entity. It is possible that ADEM might represent the initial episode of symptomatic form of what later may or may not evolve into a subtype of MS in which T cells play a significant role in demyelination. It is also conceivable that perivenular demyelination and confluent demyelination may have same pathologic basis and differ only in the stage and the appearance of the lesions. In this regard, it is of particular interest that clustering of CD8+ lymphocytes as seen in the presented case has also been described in MS and the pathogenetic role of their clonal expansion has been emphasized [11]. One may speculate that such clonal cell expansion in perivenous demyelination may act as a focus of further demyelination, resulting in confluent demyelination.

It is unknown what factor(s) may be responsible to evolution of ADEM into MS. Differences in the inherent immunological factors of the individual patient may contribute to the development of various types of MS [25]. Mowry et al. have shown that patients with monofocal lesion on MRI at the time of clinically isolated syndrome (CIS) were associated with higher risk of second relapse (i.e., clinically definite MS) than multifocal patients [26]. They suggested that patients who have more concurrent or destructive demyelinating lesions are more prone to temporarily suppress biologic disease processes compared with those with less aggressive disease onset. Van der Valk et al. suggested that existence of some level of intrinsic regulation may stop lesion progression at an early stage of microglia activation and clustering [15]. On the other hand, it has been recently shown that cortical demyelination may precede demyelination in the white matter in MS [27, 28]. Popescu et al. reported on a 33 year old patient who started to complain of headache one week after an upper respiratory tract infection. She was found to have a solitary lesion in the occipital cortex and a biopsy demonstrated cortical demyelination associated perivascular infiltrates consisting of T and B cells. The diagnosis of MS was subsequently confirmed through clinical follow up and by the new appearance of white matter lesions on MRI [29]. Thus, the putative acquired mechanisms to suppress relapse of demyelination in ADEM may include factors which alter the functions of immune effector cells such as microglia and T and B lymphocytes, or those affecting entry of the activated immune cells from the subarachnoid space into brain parenchyma through BBB.

PERIPHERAL NERVOUS SYSTEM (PNS) INVOLVEMENT IN ADEM

There have been a few clinically and pathologically well-documented cases of combined ADEM and Guillain-Barre syndrome. Kinoshita et al. reported a 68 year-old man who, after a flu-like episode, developed severe disseminatied encephalomyelitis culminating in death in 10 days. The autopsy revealed disseminated perivascular demyelination in the CNS and in the spinal roots as manifested by the presence of denuded or thinly myelinated axons by electron microscopy [29]. Aimoto et al. described a 41 year-old man who developed demyelinating peripheral neuropathy during the course of ADEM following a flu-like illness. A teased nerve fiber study of the sural nerve revealed that more than 50% of the fibers showed segmental demyelination and remyelination (condition C, D, F per Dyck classification) [30]. Lassmann et al. reported a case of simultaneous inflammatory demyelination in the CNS and PNS of a 26 year-old woman who died 12 weeks after the first symptoms with acute bilateral swelling of lids and lips. The autopsy revealed typical picture of MS in the CNS and segmental demyelination in the PNS revealed by a teased nerve fiber study and ultrastructure [30].

Combined ADEM and GBS cases without biopsy or autopsy documentation have often been reported [32-38]. Nadkarni reported a 28 year-old woman who, after flu-like episode, presented with typical GBS followed by bilateral optic neuritis and MRI evidence of ADEM, two discrete events separated by an interval of 3 weeks. Oligoclonal bands were present in the CSF. There was a good response to high dose methylprednisolone. It is of interest that the development of ADEM occurred after the clinical improvement of limb weakness due to GBS following plasma exchange, and, had the clinical picture occurred in the reverse sequence GBS would have been missed [38].

In a series of 60 adult patients clinically diagnosed as ADEM, peripheral nerve involvement can be present in up to 43% of the cases [39] and it is a risk factor for steroid resistance and poor outcomes [35, 40]. The presence of anti-galactocerebroside antibody has been reported in resistant cases of ADEM with PNS involvement [41]. It is also of noteworthy that a case of ADEM with central and peripheral conduction block has been shown to improve with ultra-high dose methylpredonisolone [42].

It is quite possible that PNS lesions in ADEM may be masked by various CNS symptoms of ADEM and have only rarely been studied with nerve conduction studies, resulting in the under-recognition of this condition as has been suggested [29, 31]. The occurrence of combined demyelination of CNS and PNS may suggest the presence of shared pathogenetic CNS and PNS epitopes or a common autoimmune etiology in susceptible individuals following certain infections or vaccinations, a concept originally proposed by Blenow [41].

CONCLUSION

There has been a classical question if ADEM can be included as a part of the spectrum of MS based on reported occurrence of patients of ADEM who develop clinically definite MS later in their courses. Pathologically, ADEM and MS have been considered as distinct phenotypes characterized by perivenous demyelination vs. confluent demyelination. Moreover, unlike ADEM which shows generally homogenous pathological features of inflammatory demyelination with known similarities to experimental autoimmune encephalitis, MS shows quite diverse heterogenous pathologic patterns. However, cases of the first clinical episode of CNS demyelination showing both features of ADEM and MS do exist. Such cases may share common patterns of immune effector cell infiltration including both CD8+ T cell and B cells with activation of microglia in the lesions, in addition to both perivenous and confluent demyelination patterns. It seems, therefore, very likely that the CNS pathology of ADEM may share common pathologic mechanism(s) with certain subgroups of MS. Moreover, cases of ADEM with concomitant peripheral nerve involvement have been documented clinically and pathologically, which may suggest common underlying immune etiologies involving both the CNS and PNS.

ACKNOWLEDGMENTS

I would like to thank Ms. Emiko Chino, Department of clinical laboratory for technical assistance and Prof. Robert E. Schmidt, Professor of Neuropathology, Washington University School of Medicine, St. Louis, for reviewing the manuscript.

REFERENCES

[1] Stankiewicz J. Multiple Sclerosis and Allied demyelinating diseases. In: Ropper AH and Samuels MA and Klein JP (Eds). *Adams and Victor's Principles of Neurology* 10th ed. McGraw-Hill; 2014; pp 937-941.

[2] Moore CRW, Stadelmann-Nessler C. Acute disseminated encephalomyelitis and related disorders. In: *Greenfield Neuropathology* 9th ed. CRC Press 2015 pp 1379-1384.

[3] Young NP, Weinshenker BG, Parisi JE, et al. Perivenous demyelination: associated with clinically defined acute disseminated encephalomyelitis and comparison with pathologically confirmed multiple sclerosis. *Brain* 2010; 133; 333-348.

[4] Hartung HP, Grossman RI. ADEM; distinct disease or part of MS spectrum? *Neurology* 2001; 56: 1257-1260.

[5] Koshihara H, Oguchi K, Takei Y, et al. Meningeal inflammation and demyelination in a patient clinically diagnoses as acute disseminated encephalomyelitis. *J Neurol Sci* 2014; 346: 323-327.

[6] Kepes JJ. Large focal tumor-like demyelinating lesions of the brain: intermediate entity between multiple sclerosis and acute disseminate encephalomyelitis? A study of 31 patients. *Ann Neurol* 1993; 33: 18-27.

[7] Lucchinetti CF, Gavrilova RH, Metz I et al. Clinical and radiographic spectrum of pathologically confirmed tumefactive multiple sclerosis. Brain 2008; 131: 1759-1775.

[8] Booss J, Esiri MM, Tourtelotte WW et al. Immunohistological analysis of T lymphocyte subsets in the central nervous system in chronic multiple sclerosis. *J Neurol Sci* 1983; 62:219-32.

[9] Hauser SL, Bhan AK, Gilles F et al. Immunohistochemical analysis of the cellular infiltrate in multiple sclerosis lesions. *Ann Neurol* 198; 19:578-87.

[10] Babbe H, Roers A, Waisman A et al. Clonal expansion of CD8+ T cells dominate the T cell infiltrate in active multiple sclerosis lesions as shown by micromanipulation and single cell polymerase chain reaction. *J Exp Med* 2000; 192: 393-404.

[11] Fisher JM, Bramow S, Dal Bianco A, et al. The relation between inflammation and neurodegeneration in multiple sclerosis brains. *Brain* 2009; 132: 1175-89.

[12] Kuhlmann T, Lassmann H, Bruck W. Diagnosis of inflammatory demyelination in biopsy specimens; a practical approach. *Acta Neuropathol* 2008; 115: 275-287.

[13] Schirmer L, Seifert CL, Pfeifenbring S, et al. Clinicopathological considerations in acute disseminated encephalomyelitis (ADEM): a fulminant case with favorable outcome. *J Neurol* 2012; 259: 753-755.

[14] Van Horssen J, Singh S, van der Pol S et al. Clusters of activated microglia in normal-appearing white matter show signs of innate immune activation. *J Neuroinflammation* 2012 9:116

[15] Van der Valk P, Amor S. Preactive lesions in multiple sclerosis. Curr Opinion in Neurol 2009; 22: 207-213.

[16] Singh S, Metz I, Amor S, et al. Microglial nodules in early multiple sclerosis white matter are associated with degenerating axons. *Acta Neuropathologica*. 2013; 125: 595-608.

[17] Lucchinetti C, Bruck W, Parisi J, et al. Heterogeneity of multiple sclerosis lesions; implications for the pathogenesis of demyelination. *Ann Neurol* 2000; 47: 707-17.

[18] Schwarz S, Mohr A, Knauth M, et al. Acute disseminated encephalomyelitis: a follow-up study of 40 adult patients. *Neurology* 2001; 56: 1313-1318.

[19] de Seze, Debouverie M, Zephir H, e al. Acute fulminant demyelinating disease: a descriptive study of 60 patients. *Ach Neurol* 2007; 64: 1426-1432.

[20] Dale RC, Branson JA. Acute disseminated encephalomyelitis or multiple sclerosis: can the initial presentation help in establishing a correct diagnosis? *Arch Dis Chil* 2005; 90: 636-9.

[21] Tennembaum S, Chitnis T, Ness J, et al. Acute disseminated encephalomyelitis. *Neurology* 2007; 68(Suppl 2): S23-S36.

[22] Hoche F, Pfeifenbring S, Vlaho S, et al. Rare brain biopsy findings in a first ADEM-like event of pediatric MS: histopathologic, enuroradiologic and clinical features. *J Neural Transm* 2011; 118: 1311-1317.

[23]Guenther AD, Munoz DG. Plaque-like demyelination in acute disseminated encephalomyelitis (ADEM) – an autopsy case report. *Clin Neuropathol* 2013; 32: 486-491.

[24] Yidiz O, Pul R, Raab P et al. Acute hemorrhagic leukoencephalitis (Weston-Hurst syndrome) in a patient with relapse-remitting multiple sclerosis. *J Neuroinflammation* 2015; 12:175 DOI 10.1186/s12974-015-0398-1

[25] Metz I, Weigend SD, Popescu BFGet al. Pathologic heterogeneity persists in early active multiple sclerosis lesions. *Ann Neurol* 2014; 75: 728-738.

[26] Mowry EM, Pesic M, Grimes B et al. Clinical predictors of early second event in patients with clinically isolated syndrome. *J Neurol* 2009; 256: 1061-1066.

[27] Popescu BFGh, Bunyan RF, Parisi JE, et al. A case of multiple sclerosis presenting with inflammatory cortical demyelination. *Neurology* 2011; 76: 1705-1710.

[28] Lucchinetti CF, Popescu BFG, Bunyan RF et al. Inflammatory cortical demyelination in early multiple sclerosis. *New Engl J Med* 2011; 365: 2188-97.

[29] Kinoshita A, Hayashi M, Miyamoto M, Oda M, Tanabe H. Inflammatory demyelinating polyradiculitis in a patient with acute disseminated encephalomyelitis (ADEM). J Neurol Neurosurg Psychiatry 1996; 60: 87-90.

[30] Aimoto Y, Morikawa F, Matsumo A et al. A case of acute disseminated encephalomyelitis (ADEM) associated with demyelinatin peripheral neuropathy. *No To Shinkei* (Brain Nerve) 1996; 48: 857-860. In Japanese, abstract in English.

[31] Lassman H, Budka H, Schnaberth G. Inflammatory demyelinating polyradiculitis in a patient with multiple sclerosis. *Arch Neurol* 1981; 38: 99-102.

[32] Katchanov J, Lünemann JD, Masuhr F, et al. Acute combined central and peripheral inflammatory demyelination. *J Neurol Neurosurg Psychiatry.* 2004; 75:1784-6.

[33] Shiraiwa N, Yoshizawa T, Ohkoshi N et al. A case of acute disseminated encephalomyelitis (ADEM) associated with peripheral neuropathy. *Clin Neurol* 2007; 47: 169-172. In Japanese, abstract in English.

[34] Amit R, Glick B, Itzchak Y et al. Acute severe combined demyelination. *Childs Nerv Syst* 1992; 8: 354-356.

[35] Shahar E, Andraus J, Savitzki D et al. Outcome of severe encephalomyelitis in children: Effect of high-dose methylpredonisolone and immunogloblins. *J Child Neurol* 2002;17: 810-14.

[36] Gamstrorp I. Encephalo-myelo-radiculo-neuropathy: involvement of the CNS in children with Guillain-Barre syndrome. *Develop Med Child Neurol* 1974; 16: 654-658.

[37] Wendt JS, Burks JS. An unusual case of encephalomyelo-radiculoneuropathy in a young woman. *Arch Neurol* 1981; 38: 726-727.

[38] Nadkarni N, Lisak RP. Guillain-Barre syndrome (GBS) with bilateral optic neuritis and central white matter disease. *Neurology* 1993; 43: 842-843.

[39] Marchioni E, Ravaglia S, Piccolo G, et al. Postinfectious inflammatory disorders: subgroups based on prospective follow-up. *Neurology* 2005; 65: 1057-65

[40] Ravaglia S, Piccolo G, Ceroni M et al. Severe steroid resistant post-infectious encephalomyelitis: general features and effects of IVIG. *J Neurol* 2007; 254: 1518-23.

[41] Samukawa M, Hirano M, Tsugawa M et al. Refractory acute disseminated encephalomyelitis with anti-galactocerebroside antibody. *Neurosci Res* 2012; 74: 284-289.

[42] Uehara A, Okamoto Y, Nishitarumi K et al. A patient of ADEM with central and peripheral conduction block improved with ultra-high-dose methylpredonisolone. *Clin Neurol* 2002; 42: 237-239. In Japanese, abstract in Emglish.

[43] Brennow G, Gamstorp I, Rosenberg R. Encephalo-myelo-radiculo-neuropathy. *Develop Med Child Neurol* 1968; 10: 485-490.

In: Horizons in Neuroscience Research. Vol. 25   ISBN: 978-1-63485-286-9
Editors: A. Costa and E. Villalba   © 2016 Nova Science Publishers, Inc.

Chapter 7

"MENINGITIS–RETENTION SYNDROME" AS A MILD FORM OF ACUTE DISSEMINATED ENCEPHALOMYELITIS

Ryuji Sakakibara[1], MD, PhD, Fuyuki Tateno[1],
Masahiko Kishi[1], Yohei Tsuyusaki[1], Yosuke Aiba[1],
Hiromi Tateno[1], Tsuyoshi Ogata[1], Tatsuya Yamamoto[2],
Tomoyuki Uchiyama[3] and Tomonori Yamanishi[3]*

[1]Neurology, Internal Medicine,
Sakura Medical Center,
Toho University, Sakura, Japan
[2]Neurology, Chiba University, Chiba, Japan
[3]Continence Center, Dokkyo Medical College,
Tochigi, Japan

ABSTRACT

Aims: A peculiar combination of acute urinary retention and aseptic meningitis has been described. This combination is referred to as meningitis–retention syndrome (MRS), since patients with this syndrome exhibited no other abnormalities, except for mild pyramidal involvement. We aimed to delineate this syndrome by reviewing literatures.

* E-mail: sakakibara@sakura.med.toho-u.ac.jp. Phone: +81-43-462-8811 ext.2323, Fax: +81-43-487-4246.

Methods: We performed a systematic review of the literature to identify the frequency, clinical symptoms, urodynamic findings, putrative underlying pathology, and management of this syndrome.

Results: Patients with MRS have typical symptoms of fever, headache, stiff neck, and minor pyramidal signs, together with acute urinary retention. The bladder is initially areflexic, but soon becomes either normal or overactive in the repeated urodynamics during the course of the disorder. MRS is thought to be a very mild form of acute disseminated encephalomyelopathy (ADEM), with increased cell count, total protein, and occasional myelin basic protein in the cerebrospinal fluid. Proper management of the acute urinary retention is necessary to avoid bladder injury due to overdistension. The effectiveness of immune treatments (e.g., steroid pulse therapy) in shortening the urinary retention period awaits further study.

Conclusions: Although rare, MRS is thought to be a very mild form of ADEM, and a disorder that both urologists and neurologists may encounter. MRS should be listed in the differential diagnosis of acute urinary retention.

Keywords: meningitis-retention   syndrome,   acute   disseminated encephalomyelopathy, urinary retention

INTRODUCTION

Urinary retention is a urologic emergency. Among patients with acute urinary retention, children, young adults, and women are extremely rare [1]. Cases of urinary retention in such occasion may have a neurologic etiology [1]. Spina bifida occulta / tethered cord syndrome is one disorder known to lead to urinary retention without marked neurological abnormalities, except for saddle anesthesia [2]. Fowler's syndrome is another disorder that causes urinary retention, particularly in neurologically intact young women; isolated urethral sphincter hypertonicity underlies this condition [3]. Inflammatory neurologic diseases also cause acute urinary retention, in which patients lack apparent urethral outlet obstruction, but exhibit only minor neurological as well as cerebrospinal fluid (CSF) abnormalities. Based on the mechanism of urinary retention, these disorders can be divided into two subgroups: disorders of the peripheral nervous system (e.g., sacral herpes, caused by herpes simplex virus or varicella zoster virus, with unilateral sacral pain, sensory signs and often skin rashes in the same area [4]); and disorders of the central nervous system (e.g., meningitis–retention syndrome (MRS), with fever, headache, stiff neck, and minor pyramidal signs). Previously, the latter condition has been only occasionlly reported, but now it is recognized as a specific category [5, 6]. Recently, there are evidences suggesting that MRS is thought to be a very mild form of acute disseminated encephalomyelopathy (ADEM). We here performed a systematic review of the literature to identify the frequency, clinical symptoms, urodynamic findings, putrative underlying pathology, and management of this syndrome.

PATHOPHYSIOLOGY

Summary of Three Individual Cases

In 2005, Sakakibara and co-workers [5] reported three cases of a peculiar combination of acute urinary retention and aseptic meningitis that the authors referred to as MRS, since these cases exhibited no other abnormalities, except for mild pyramidal involvement. The mechanism of urinary retention of these cases was investigated by urodynamics studies. The patients consisted of two men and one woman (age range: 34-68 years; Tables 1, 2). All three patients developed acute urinary retention along with headache, fever, and stiff neck. None of these patients had obvious neurological abnormalities, other than a slightly brisk reflex in the lower extremities in two of the three patients. One patient previously experienced generalized erhythematous eruptions, but none had pain, hypoalgesia, or skin eruptions in the sacral dermatomes suggestive of sacral herpes. Their brain/spinal/lumbar plexus MRI scans and nerve conduction studies were all normal. CSF examination showed mild mononucleolar pleocytosis, increased protein content, and normal to mildly decreased glucose content in all patients; increased myelin basic protein (MBP) suggestive of central nervous system demyelination in one patient; and increased viral titers in none of the patients. Urodynamic study revealed voiding-phase detrusor areflexia in all patients and an unrelaxing sphincter in one patient. These clinical manifestations were ameliorated within 3 weeks. To summarize, the authors reported 3 cases with a combination of acute urinary retention and aseptic meningitis alone, with mildly brisk reflexes and increased myelin basic protein, none of whom showed cauda equina signs or symptoms. This condition was considered to be a very mild form of ADEM.

In its typical form, ADEM appears after vaccination or after exanthematous infections, and serological and pathological studies have suggested that ADEM is of parainfectious/autoimmune origin [7].

Table 1. Meningitis–retention syndrome and related conditions

disease	neurological symptoms/signs						urodynamics	cerebrospinal fluid (CSF)		MRI	viral titers
	headache fever	stiff neck Kernig	DOC convulsion	LE reflexes	motor	sensory	bladder function	increased cells, protein	MBP, OCB		
ADEM	+	+	+	N or brisk or decreased	paraparesis	sensory level or saddle	DHIC	+	+	WML & spinal cord	usually no
meningitis-retention syndrome	+	+	-	N or brisk	-	N	DA (spinal shock?) > DH	+	+-	-	usually no
myelitis	-+	-+	-	brisk	paraparesis	sensory level	DHIC	+	+-	spinal cord	usually no
sacral herpes	-	-	-	N or decreased	-	unilateral saddle, skin erruption	DA	+	-	-	VZV, HSV

ADEM: acute disseminated encephalomyelitis.
DOC: disturbance of consciousness.
LE: lower extremities.
N: normal.
Sensory level: sensory disturbance below the cervical/thoracic level.
Saddle: sensory disturbance in the saddle/sacral area.
DHIC: detrusor hyperreflexia with impaired contractile function (see text).
DA: detrusor areflexia.
MBP: myelin basic protein.
OCB: oligoclonal band.
WML: white matter lesion.
HSV: herpes simplex virus.
VZV: varicella zoster virus.

Antecedent/ comorbid infections are diverse, include human herpes virus 6 [8]. A combination of signs of encephalitis (e.g., disturbance of consciousness, epilepsy, and hemiparesis), signs of myelitis (e.g., sensory disturbance below the level of the lesion, spastic paraplegia, whereby flaccid parapalegia may occur in the initial shock stage), and lower urinary tract (LUT)/bowel dysfunction typically occur in patients with ADEM. Lesions observed in brain MRI are usually confined to the white matter. Lesions in the spinal cord involving the conus are also visualized [6]. CSF pleocytosis, increased protein, mildly decreased glucose content, increased MBP/oligoclonal bands, and a lack of increased viral titers, are all features of this disease. Patients with ADEM commonly exhibit LUT dysfunction [9, 10], which varies from urinary retention to urgency incontinence. LUT dysfunction appears to be related to pyramidal tract involvement, and most probably reflects the severity of the spinal cord lesion.

Urodynamic findings have included detrusor overactivity in the storage phase, brisk bulbocavernosus reflex, and detrusor-sphincter dyssynergia (reflective of a suprasacral spinal cord lesion); detrusor areflexia in the voiding phase (reflective of either acute spinal shock or conus lesion); and neurogenic sphincter EMG in one of four patients studied (reflective of a conus lesion) [9, 10]. The responsible lesion sites in cases of LUT dysfunction have appeared to be the cervico-thoracic spinal cord, and, to a lesser extent, the conus [11] (and possibly the spinal nerve roots [12]) in patients with ADEM. In addition, cases of ADEM/ parainfectious myelitis presented with LUT dysfunction alone, either initially [13, 14], or as the only remaining consequence of the disease [15], thus suggesting that LUT innervation was selectively vulnerable in these cases. In some cases, abnormal F-waves were recorded, suggesting conus or a radicular lesion [6]. Although aseptic meningitis is a common neurological disorder, the combination of aseptic meningitis and acute urinary retention (MRS) has not been previously well recognized, and the prevalence of this disorder is not known. The clinical manifestations of Sakakibara's three cases [5] are mostly the same as those of the six reported cases in the literature (including cases written in Japanese) [16].

Table 2. Reported cases of meningitis–retention syndrome

year, author	age	sex	fever	head-ache	stiff neck/ Kernig	drowsi-ness	LE rfx	sens-ation	FS ml	BC ml	DA	US/ DSD	cell /mm3	P:M	prot mg/dl	glu mg/dl CSF:serum	MBP	brain	spinal cord	meningitis (weeks)	retention (weeks)
1985 Kanno	34	F	+	+	+		-	brisk	numb		460	+		40			44:94 (47%)				3	6
1990 Ohe	24	F	+	+	-		-	N	N	170	210	+?	+	312	0:100	260			N		2	2
1996 Fukagai	46	M	+	-	-	+		N	N	96	220	+?		143	19:81						6	10
1999 Shimizu	13	M	+	+	+		-	N	N	N	N	+		60	1:99	N	N				4	5
	18	F	+	+	+		-	N	N	N	N	+		109	34:66	76	N		N	N	3	4
2005 Sakaki-bara	46	M	+	+	+		-	brisk	N	130	500	+	+	290	2:98	80	41:125 (33%)	-	N	N	2	3
	68	F	+	+	+-	+-		N	N	180	280	+		108	5:95	97	41:92 (45%)	+	N	N	2	3
	34	M	+	+	-			brisk	N	190	460	+	-	38	0:100	71	57:93 (61%)		N	N	1	2

Modified from Sakakibara et al. 2005.

Kernig: Kernig sign.

LE: lower extremities.

Rfx: reflexes.

FS: first sensation (100 < normal < 300).

BC: bladder capacity (200 < normal < 600).

DA: detrusor areflexia.

US: unrelaxing sphincter.

DSD: detrusor-sphincter dyssynergia.

P:M: polymorphonuclear cells: mononuclear cells.

Prot: protein.

Glu: glucose.

N: normal.

As Table 2 shows, Kanno and colleagues are the group who first reported MRS. The clinical manifestations of MRS cases differ markedly from those of ADEM, because other than presenting with aseptic meningitis, MRS cases lack apparent encephalitic signs such as disturbance of consciousness, epilepsy or aphasia, and myelitic signs such as gait abnormalities, and sensory-level abnormalities. Only one patient reported by Fukagai et al. showed drowsiness without meningeal irritation. Whereas brain and spinal cord lesions typically appear in ADEM, such change is not observed in MRS. However, recently, a reversible splenial lesion was noted in a brain MRI of a patient with MRS [17]. The CSF examination of this patient showed a mononuclear pleocytosis of 38–370 /mm^3, normal to increased protein content (up to 260 mg/dl), and normal to mildly decreased glucose content (up to 33% of that in the serum). All viral titers studied in the CSF and the serum of such cases were negative. Nevertheless, it appears likely that the urinary retention in such patients is of neurologic etiology, since none of the patients appear to exhibit urologic abnormalities such as prostatic hyperplasia, and also because a strong chronological association is observed between the onset of urinary retention with, or just after, the occurrence of the aseptic meningitis. As it is observed in ADEM, antecedent/ comorbid infections or conditions with MRS include *angiostrongylus cantonensis* [18, 19], Epstein–Barr virus [20], and herbal medicine use [21], whereas such antecedent infections are not apparent in the remaining cases [22, 23]. Urodynamic study results have shown that all patients examined had detrusor areflexia, which results in an inability to contract the bladder on voiding, and two patients had an unrelaxing sphincter (one case of Ohe and Sakakibara's case 1). Detrusor areflexia originates from various lesion sites along the neural axis; most commonly, PNS lesions are observed. However, CNS lesions that affect the spinal cord or the brain can also cause detrusor areflexia, which is particularly seen in the acute-shock phase of patients with transverse myelitis (often immune-mediated) or ADEM. Tateno et al. recently encountered a man with MRS in whom a urodynamic study was performed twice. In that case, an initially areflexic detrusor became overactive after a 4-month period, suggesting an upper motor neuron bladder dysfunction [24]. As described above, encephalitic features are absent in patients with MRS. However, three MRS patients (one case reported by Kanno et al. and Sakakibara's cases 1 and 3) showed the brisk lower-extremity reflexes suggestive of mild myelitis. In one of Sakakibara's cases, MBP was increased, which is suggestive of CNS demyelination. Therefore, it appears possible that MRS is a very mild variant of ADEM, which selectively affects LUT innervation.

DIAGNOSIS AND MANAGEMENT

Arriving at a diagnosis is not straightforward when a combination of aseptic meningitis and acute urinary retention is confirmed, and such a diagnosis should be made both urologically and neurologically (Figure 1) in order to exclude sacral herpes, a benign variant of Guillain-Barré syndrome (inflammatory radiculitis), typical ADEM, myelitis with leg weakness [25], typical multiple sclerosis (MS), neuromyelitis optica (NMO), herpetic brainstem encephalitis, chemical meningitis secondary to focal subarachnoid bleeding, and other common causes of neurogenic urinary retention (e.g., diabetic neuropathy, lumbar spondylosis, etc.). For the differential diagnosis, a detailed history, brain/spinal MRI and nerve conduction study are necessary.

In MRS, to date at what level the central nervous system pathology causes retention remains unclear. However, as described above, the brain seems less contributing to urinary retention since none of MRS cases showed encephalitic episodes or white matter lesions by MRI, except for a posterior splenial lesion [17]. In contrast, minimum spinal cord lesion seems contributing to urinary retention in MRS since repeated urodynamics showed an appearence of detrusor overactivity [24], and some cases showed brisk deep tendon reflexes (without weakness).

The term "Elsberg syndrome" has been occasionally used, which is rather vaguely assigned to sacral myeloradiculitis of undetermined etiology, sacral herpes, sacral vasculitic neuropathy, possible MRS, possible ADEM, or conus infarction. However, in contrast to the majority of "Elsberg syndrome" cases, Kennedy, Elsberg, and Lambert (1913) [26] reported five cases of pathology-demonstrated cauda equina radiculitis. Clinical pictures of these cases were characterized by rare CSF abnormalities (only one of the four cases described showed an increased cell count as well as increased protein levels); no clinical meningitis; a subacute/chronic course (in one case the course of disease was subacute, lasting only 3 months, but in the remaining four cases, the course of disease was chronic, lasting 4.5–36 months); presentation with typical cauda equina motor-sensory-autonomic syndrome (in particular, four of the patients had apparent muscle atrophy/weakness in the lumbosacral segment); Wallerian degeneration of the spinal afferent tracts; and mild upper motor neuron signs. The authors assumed that the etiology of these cases was either inflammatory or toxic.

The exact cause of these five cases remains uncertain, although clinical pictures resemble, at least in part, those of paraneoplastic [27] or autoimmune lumbosacral radiculoplexus neuropathy [28].

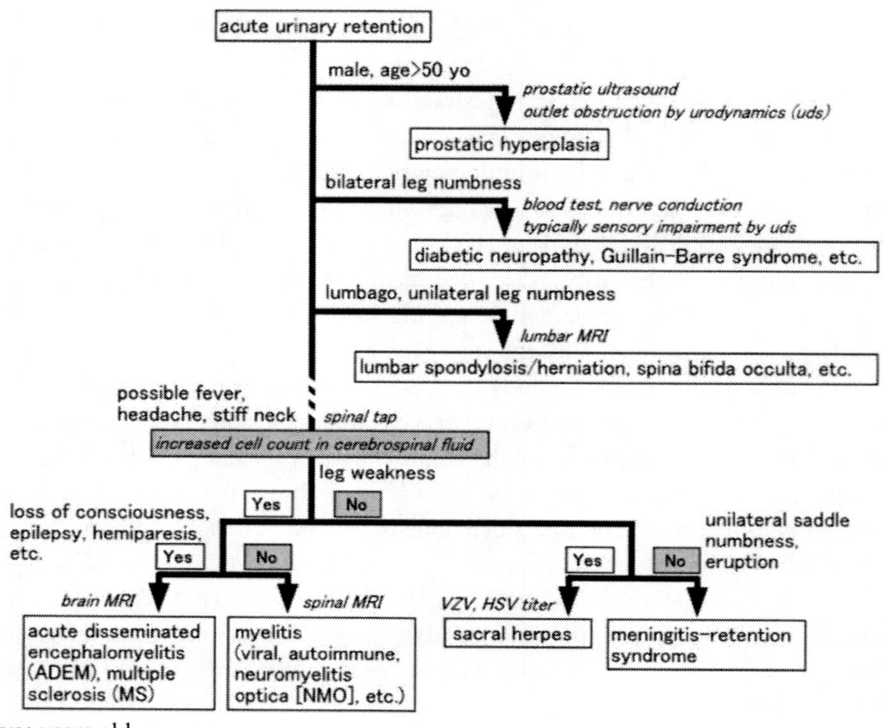

yo: years old.
MRI: magnetic resonance imaging.
HSV: herpes simplex virus.
VZV: varicella zoster virus.
Cited from ref. 29.

Figure 1. Algorithm to diagnose meningitis–retention syndrome and related conditions.

Clinical features of sacral herpes or MRS differ markedly from those of Kennedy's cases. While acute MS, ADEM and NMO need steroid pulse or extensive immune therapy, MRS has a benign and self-remitting course (i.e., a duration of 2–10 weeks), and the effectiveness of immune treatments (e.g., steroid pulse therapy) remains unclear, although such treatments may shorten the duration of the disease. Management of acute urinary retention is necessary to avoid bladder injury due to overdistension [29].

CONCLUSION

This paper reviews meningitis–retention syndrome (MRS), having typical symptoms of fever, headache, stiff neck, and minor pyramidal signs, and thought to be a very mild form of acute disseminated encephalomyelopathy (ADEM). Both urologists and neurologists may encounter this disorder. The bladder is initially areflexic, but soon becomes either normal or overactive in the repeated urodynamics during the course of the disorder. While MRS is self-remitting and has a benign course, proper management of the acute urinary retention by clean, intermittent catheterization, etc., is necessary to avoid bladder injury due to overdistension. The effectiveness of immune treatments (e.g., steroid pulse therapy) in shortening the urinary retention period in MRS remains unclear and warrants further reseach.

ACKNOWLEDGMENT

Conflict of Interest

None of the authors have conflict of interest.

Author Contributions

Ryuji Sakakibara has a role in: study concept and design, acquisition of subjects and/or data, analysis and interpretation of data, and preparation of manuscript.

Masahiko Kishi has a role in: acquisition of subjects and/or data

Akihiko Tateno has a role in: acquisition of subjects and/or data

Fuyuki Tateno has a role in: acquisition of subjects and/or data

Yohei Tsuyusaki has a role in: acquisition of subjects and/or data

Tomoyuki Uchiyama has a role in: acquisition of subjects and/or data

Tatsuya Yamamoto has a role in: acquisition of subjects and/or data

REFERENCES

[1]   Herbaut A. Neurogenic urinary retention. *Int. Urogynecol. J.,* 1993; 4: 221-8.

[2]   Sakakibara R., Hattori T., Uchiyama T., Kamura K., Yamanishi T. Uro-neurological assessment of spina bifida cystica and occulta. *Neurourol. Urodynam.,* 2003; 22(328-334).

[3]   Swinn M., Lowe E., Fowler C. The clinical features of non-psychogenic urinary retention (Fowler's syndrome). *Neurourol. Urodynam.,* 1998; 17:383-4.

[4]   Yamanishi T., Yasuda K., Sakakibara R., Hattori T., Uchiyama T., Minamide M., et al. Urinary retention due to Herpes Simplex infections. *Neurourol. Urodynam.,* 1998; 17:613-9.

[5]   Sakakibara R., Uchiyama T., Liu Z., Yamamoto T., Ito T., Uzawa A., et al. Meningitis-retention syndrome; an unrecognized clinical condition. *J. Neurol.,* 2005; 252:1495-9.

[6]   Sakakibara R., Yamanishi T., Uchiyama T., Hattori T. Acute urinary retention due to benign inflammatory nervous diseases. *J. Neurol.,* 2006 253:1103-10.

[7]   Schwarz S., Mohr A., Knauth M., Wildemann B., Storch-Hagenlocher B. Acute disseminated encephalomyelitis. A follow-up study of 40 adult patients. *Neurology,* 2001; 56:1313-8.

[8]   Kawamura M., Kaku H., Takayama N., Ushimi T., Kishida S. Acute Urinary Retention Secondary to Aseptic Meningoencephalitis in an Infant - a Case Report. *Brain Nerve,* 2007; 59:1287-91.

[9]   Sakakibara R., Hattori T., Yasuda K., Yamanishi T. Micturitional disturbance in acute disseminated encephalomyelitis (ADEM). *J. Auton. Nerv. Syst.,* 1996; 60:200-5.

[10]   Panicker J., Nagaraja D., Kovoor J., Nair K., Subbakrishna D. Lower urinary tract dysfunction in acute disseminated encephalomyelitis. *Mult. Scler.,* 2009 15:1118-22.

[11]   Pradhan S., Gupta R. K., Kapoor R., Shashank S., Kathuria M. K. Parainfectious conus myelitis. *J. Neurol. Sci.,* 1998; 161:156-62.

[12]   Kinoshita A., Kaseda S., Yagi K., Oda M., Tanabe H. A case of acute disseminated encephalomyelitis with pathology-proven acute demyelinating lesion in the peripheral nervous system. *Clin. Neurol.,* 1994; 34:892-7.

[13]   Kusuhara T., Nakajima M., Inoue H., Takahashi M., Yamada T. Parainfectious encephalomyeloradiculitis associated with herpes simplex

virus 1 DNA in cerebrospinal fluid. *Clin. Infect. Dis.,* 2002; 34:1199-205.

[14] Hiraga A., Sakakibara R., Mori M., Yamanaka Y., Ito S., Hattori T. Urinary retention can be the sole initial manifestation of acute myelitis. *J. Neurol. Sci.,* 2006 251:110-2.

[15] Hiraga A., Sakakibara R., Mori M., Suzuki A., Hattori T. Bilateral lesion in the lateral columns and complete urinary retention: association with the spinal cord descending pathway for micturition. *Neurourol. Urodynam.,* 2005; 24:398-89.

[16] Zenda T., Soma R., Muramoto H., Hayase H., Orito M., Okada T., et al. Acute urinary retention as an unusual manifestation of aseptic meningitis. *Internal Medicine,* 2002; 41:392-4.

[17] Tascilar N., Aydemir H., Emre U., Unal A., Ataso T., Ekem S. Unusual combination of reversible splenial lesion and meningitis-retention syndrome in aseptic meningomyelitis. *Clinics,* 2009; 64:932-7.

[18] Furugen M., Yamashiro S., Tamayose M., Naha Y., Miyagi K., Nakasone C., et al. Elsberg syndrome with eosinophilic meningoencephalitis caused by Angiostrongylus cantonensis. *Intern. Med.,* 2006; 45:1333-6.

[19] Hsu J., Chuang S., Chen C., Huang M. Sacral myeloradiculitis (Elsberg syndrome) secondary to eosinophilic meningitis caused by Angiostrongylus cantonensis. *BMJ Case Rep.,* 2009;pii: bcr10. 2008.1075.

[20] Ntziora F., Alevizopoulos A., Konstantopoulos K., Kanellopoulou S., Bougas D., Stravodimos K. Aseptic Meningitis with Urinary Retention: A Case Report. Case Reports in Medicine, 2011; ID 741621:3 pages.

[21] Takahashi O., Sakakibara R., Kishi M., Matsuzawa Y., Ogawa E., Sugiyama M., et al. Herbal medicine-induced meningitis-retention syndrome. *Intern. Med.,* 2010; 49:1813-6.

[22] Ito Y., Uchida Y., Tamai N., Nakajima F. Two cases of urinary retention secondary to benign inflammatory nervous diseases. *Hinyokika Kiyo,* 2009 55:655-9.

[23] Kim T., Whang J., Lee S., Choi J., Park S., Lee J. Acute Urinary Retention due to Aseptic Meningitis: Meningitis-Retention Syndrome. *Int. Neurourol. J.,* 2010 14:122-4.

[24] Tateno F., Sakakibara R., Sugiyama M., Takahashi O., Kishi M., Ogawa E., et al. Meningitis-retention syndrome: first case of urodynamic follow-up. *Intern. Med.,* 2011; 50:1329-32.

[25] Fujita K., Tanaka T., Kono S., Narai H., Omori N., Manabe Y., et al. Urinary Retention Secondary to Listeria Meningitis. *Inter. Med.,* 2008; 47:1129-31.

[26] Kennedy F., Elsberg C. A., Lambert C. I. A peculiar undescribed disease of the nerves of the cauda equina. *Am. J. Med. Sci.,* 1913; 147:645-67.

[27] Burton M., Anslow P., Gray W., Donaghy M. Selective hypertrophy of the cauda equina nerve roots. *J. Neurol.,* 2002; 249:337-40.

[28] Dyck P., Windebank A. Diabetic and nondiabetic lumbosacral radiculoplexus neuropathies: new insights into pathophysiology and treatment. *Muscle Nerve,* 2002; 25:477-91.

[29] Sakakibara R., Kishi M., Tsuyusaki Y., Tateno A., Tateno F., Uchiyama T., Yamamoto T., Yamanishi T., Yano M. "Meningitis-retention syndrome": a review. *Neurourol. Urodyn.,* 2013; 32: 19-23.

BIOGRAPHICAL SKETCH

Dr. Ryuji Sakakibara was born and raised in Japan. He received his medical degree from the Asahikawa Medical College in 1984. Because of interest in clinical neurology, he moved forward to Chiba University (Professor Keizo Hirayama) where he subsequently worked as an assistant Professor, and a Lecturer in the Department of Neurology.

He completed his PhD at the Department of Neurology, Chiba University, in 1992 (Professor Takamichi hattori). From 1997 to 1998 he was a research fellow in the National Hospital for Neurology and Neurosurgery / Institute of

Neurology, Queen Square, London, supervised by Professor Clare J. Fowler (retired in 2010) in the Department of Uro-Neurology. In 2007 March, he started a new Department of Neurology in Toho University, Sakura Medical Center, Sakura, Japan as an Associate Professor. He is a member of many international societies including the American Academy of Neurology, the Association of British Neurologists, the Movement Disorders Society, the International Society for Autonomic Neuroscience, the European Federation of Autonomic Societies, the American Autonomic Society, the International Continence Society, and the International Consultation on Incontinence. He also serves as an adhoc reviewer of several journals including Neuroscience Letters, Journal of Neurology, Neurosurgery and Psychiatry, Journal of Neurology, European Journal of Neurology, Movement Disorders, Muscle and Nerve, Urology, Journal of Urology, Clinical Physiology, and American Journal of Physiological Medicine. During his career, Dr. Sakakibara has authored more than 180 articles in peer-reviewed international journals.

SYNOPSIS OF AREA OF INTEREST: Dr. Sakakibara's research interests include neuro-urology, neuro-gastroenterology, autonomic physiology, drug trials and clinical neurology including Parkinson's disease and multiple system atrophy.

CURRICULUM VITAE

Ryuji Sakakibara, PhD
Associate Professor, Neurology Division, Department of Internal Medicine
Sakura Medical Center, Toho University
564-1 Shimoshizu, Sakura, 285-8741 Japan
Telephone: +81-43-462-8811 extension 2323
Fax: +81-43-487-4246
e-mail: sakakibara@sakura.med.toho-u.ac.jp
Website: http://www.lab.toho-u.ac.jp/med/sakura/neurology/ (in Japanese)

Date of Birth: 20th September, 1959
Place of birth: Shimizu, Japan
Nationality: Japanese

Education:

1979-1984	Faculty of Medicine, Asahikawa Medical College Awarded the degree of MD.
1984-1992	Department of Neurology, Chiba University School of Medicine Awarded the degree of PhD in Neurology for a thesis entitled "Pathophysiology of micturitional disturbance in patients with multiple system atrophy". Work supervised by Professor Keizo Hirayama in the Department of Neurology (retired).
1988-present	Certified Neurologist by Japan Neurologist Association Board (No. 1093)

Research and professional experience:

1984-1993	Member of the Department of Neurology, Chiba University School of Medicine
1993-1995	Chief of the Department of Neurology, Kashima Rosai Hospital
1996-2002	Assistant Professor in the Department of Neurology, Chiba University School of Medicine
1997-1998 May	Research fellow in the National Hospital for Neurology and Neurosurgery / Institute of Neurology, Queen Square, London, supervised by Dr. Clare J. Fowler in Uro-Neurology.
2003-2007 February	Lecturer in the Department of Neurology, Chiba University School of Medicine, working with Professor Takamichi Hattori (retired).
2007 March	Associate Professor and Chief in the Department of Neurology in Toho University, Sakura Medical Center, Sakura, Japan.

Membership of academic societies:

International Continence Society (Committee Member)
International Consultation in Incontinence (Committee Member)
American Autonomic Society
International Society for Autonomic Neuroscience
European Federation of Autonomic Societies
Movement Disorders Society
International Medical Society of Paraplegia

British Association of Neurologists
American Academy of Neurology
Japanese Continence Society (Senior Board Member)
Japan Medical Society of Paraplegia (Senior Board Member)
Japanese Society of Autonomic Nervous System (Board Member)
Japanese Society of Neurological Therapeutics (Board Member)
Japanese Society of Geriatric Urology (Board Member)
Japanese Society of Neurology (Board Member)

Membership of journal editorials:
Scientific World Journal (Editorial Board Member)
Neurology International (Editorial Board Member)
American Journal of Neurodegenerative Diseases (Senior Editorial Board)
Journal of Neurological Disorders and Stroke (Editorial Board Member)
Case Reports in Urology (Editorial Board Member)
LUTS (Editorial Board Member)
Bladder (Editorial Board Member)

Ad Hoc Reviewer of journals:
Neuroscience Letters, Journal of Neurology, Neurosurgery and Psychiatry, Journal of Neurology, European Journal of Neurology, Muscle and Nerve, Urology, Journal of Urology, Clinical Physiology, American Journal of Physiological Medicine, etc.

Awards:
1996 Travel award to attend International Continence Society 26th Annual Meeting in Athens, by Nakamura Scholarship.

1997 Scholarship by Japan Foundation for Aging and Health, for the research of neurophysiological study of geriatric urinary incontinence.

1998 Scholarship by Ministry of Education in Japan, for the research of neurophysiological study of geriatric micturitional disturbance.

Publications:

A. Papers

[1]   Hattori T., Yasuda K., Sakakibara R., Yamanishi T., Kitahara H., Hirayama K.: Micturitional disturbance in ossification of the posterior longitudinal ligament in the cervical spine. *J. Spinal Disorders*, 3: 4; 285-287 1990.

[2]   Hattori T., Sakakibara R., Yasuda K., Murayama N., Hirayama K.: Micturitional disturbance in cervical spondylotic myelopathy. *J. Spinal Disorders*, 3: 1; 16-18 1990.

[3]   Hattori T., Yasuda K., Sakakibara R., Yamanishi T., Kitahara H., Hirayama K.: Micturitional disturbance in tumors of the lumbosacral area, *J.Spinal Disorders*, 5: 2; 193-197 1992.

[4]   Sakakibara R., Hattori T., Tojo M., Yamanishi T., Yasuda K., Hirayama K.: Micturitional disturbance in multiple system atrophy. *Jpn J. Psychiat. Neurol.,* 47: 3; 591-598 1993.

[5]   Sakakibara R., Hattori T., Tojo M., Yamanishi, T., Yasuda K., Hirayama K.: Micturitional disturbance in progressive supranuclear palsy. *J. Autonom. Nerv. Syst.*, 45: 101-106 1993.

[6]   Sakakibara R., Hattori T., Tojo M., Yamanishi T., Yasuda K., Hirayama K.: Micturitional disturbance in radiation myelopathy. *J. Spinal Disorders,* 6: 402-405 1993.

[7]   Yasuda K., Yamanishi T., Hattori T., Murayama N., Sakakibara R., Shimazaki J.: Lower urinary tract dysfunction in the anterior spinal artery syndrome. *J. Urol.*, 150: 1182-1184 1993.

[8]   Fukutake T., Hirayama K., Sakakibara R.: Contralateral selective saccadic palsy after a small hematoma in the corona radiata ajacent to the genu of the internal capsule. *J. Neurol. Neurosurg. Psychiatry*, 56: 221 1993.

[9]   Yamanishi T., Yasuda K., Tojo M., Hattori T., Sakakibara R., Shimazaki J.: Effect of beta2-stimulants on contractility and fatigue of canine urethral sphincter. *J. Urol.*, 151: 1066-1069 1994.

[10]   Yamanishi T., Yasuda K., Tojo M., Hattori T., Sakakibara R., Shimazaki J.: Improvement of urethral resistence after administration of an alpha adrenocepter blocking agent, urapidil, for neuropathic voiding dysfunction. *Paraplegia,* 32: 271-276 1994.

[11] Hattori T., Sakakibara R., Yamanishi T., Yasuda K., Hirayama K.: Micturitional disturbance in human T-lymphotropic virus type-1-associated myelopathy. *J. Spinal Disorders*, 7: 3; 255-258 1994.

[12] Sakakibara R., Hattori T., Tojo M., Yamanishi T., Yasuda K., Hirayama K.: The location of the paths subserving micturition; studies in patients with cervical myelopathy. *J. Autonom. Nerv. Syst.*, 55: 165-168 1995.

[13] Sakakibara R., Hattori T., Tojo M., Yamanishi T., Yasuda K., Hirayama K.: Micturitional disturbance in myotonic dystrophy. *J. Autonom. Nerv. Syst.*, 52: 17-21 1995.

[14] Sakakibara R., Fukutake T., Kita K., Hattori T.: Treatment of erythromelalgia with cyproheptadine. *J. Autonom. Nerv. Syst.*, 58: 121-122 1996.

[15] Sakakibara R., Hattori T., Yasuda K., Yamanishi T.: Micturitional disturbance after acute hemispheric stroke; analysis of the lesion site by CT and MRI. *J. Neurol. Sci.*, 137: 47-56 1996.

[16] Sakakibara R., Hattori T., Yasuda K., Yamanishi T. Micturitional disturbance in acute disseminated encephalomyelitis (ADEM). *J. Auton. Nerv. Syst.*, 60: 200-205 1996.

[17] Sakakibara R., Hattori T., Yasuda K., Yamanishi T.: Micturitional disturbance and pontine tegmental lesion; urodynamic and MRI analyses of the vascular cases. *J. Neurol. Sci.*, 141: 105-110 1996.

[18] Sakakibara R., Hattori T., Yasuda K., Yamanishi T.: Micturitional disturbance in acute transverse myelitis. *Spinal Cord*, 34: 481-485 1996.

[19] Sakakibara R., Hattori T., Yasuda K., Yamanishi T.: Micturitional disturbance in syringomyelia. *J. Neurol. Sci.*, 143: 100-106 1996.

[20] Fukutake T., Kita K., Sakakibara R., Takagi K., Tokumaru Y., Kojima S., Hattori T., Hirayama K.: Late onset hereditary ataxia with global thermoanalgesia and absence of fungiform papillae on the tongue in a Japanese family. *Brain,* 119: 1011-1021 1996.

[21] Sakakibara R., Hattori T., Yasuda K., Yamanishi T., Tojo M., Mori M.: Micturitional disturbance in Wernicke's encephalopathy. *Neurourol. Urodynam.*, 16: 111-115 1997.

[22] Sakakibara R., Hattori T., Kuwabara S., Yamanishi T., Yasuda K.: Micturitional disturbance in patients with Guillain-Barre syndrome. *J. Neurol. Neurosurg. Psychiatry,* 63: 5; 649-653 1997.

[23] Sakakibara R., Hattori T., Kita K., Yamanishi T., Yasuda K.: Urodynamic and cardiovascular measurements in patients with micturition syncope. *Clin. Autonom. Res.*, 7: 219-221 1997.

[24] Yamanishi T., Yasuda K., Sakakibara R., Hattori T., Ito H., Murakami S.: Pelvic floor electrical stimulation in the treatment of stress incontinence; an investigational study and a placebo-controled double-blind trial. *J. Urol.*, 158: 2127-2131 1997.

[25] Yamanishi T., Yasuda K., Sakakibara R., Hattori T., Tojo M., Ito H.: The nature of detrusor bladder neck dyssynergia in non-neurologic bladder dysfunction. *J. Auton. Nerv. Syst.,* 66: 163-168 1997.

[26] Sakakibara R., Mori M., Fukutake T., Kita K., Hattori T.: Orthostatic hypotension in a case with multiple sclerosis. *Clin. Autonom. Res.*, 7: 163-165 1997.

[27] Fukutake T., Sakakibara R., Mori M., Araki M., Hattori T.: Chronic intractable headache in a patient with Marfan's syndrome. *Headache*, 37: 291-295 1997.

[28] Sakakibara R., Hattori T., Kita K., Arai K., Yamanishi T., Yasuda K.: Stress-induced urinary incontinence in patients with spinocerebellar degeneration. *J. Neurol. Neurosurg. Psychiatry*, 64: 3; 389-391 1998.

[29] Sakakibara R., Hattori T., Fukutake T., Mori M., Yamanishi T., Yasuda K.: Micturitional disturbance in herpetic brainstem encephalitis; contribution of the pontine micturition center (PMC), *J. Neurol. Neurosurg. Psychiatry,* 64: 2; 269-272 1998.

[30] Sakakibara R., Hattori T., Kuwabara S., Yamanishi T., Yasuda K.: Micturitional disturbance in patients with chronic inflammatory demyelinating polyneuropathy (CIDP). *Neurology*, 50: 1179-1182 1998.

[31] Yamanishi T., Yasuda K., Sakakibara R., Murayama N., Hattori T., Ito T.: Detrusor overactivity and penile erection in patients with lower lumbar spine lesions. *Eur. Urol.*, 34: 360-364 1998.

[32] Yamanishi T., Yasuda K., Sakakibara R., Hattori T., Uchiyama T., Minamide M., Ito T.: Urinary retention due to Herpes Simplex infections. *Neurourol. Urodynam.*, 17: 613-619 1998.

[33] Sakakibara R., Hattori T., Fukutake T., Mori M., Yamanishi T., Yasuda K.: Micturitional disturbance in a patient with adrenomyeloneuropathy (AMN). *Neurourol. Urodynam.*, 17: 207-212 1998.

[34] Sakakibara R., Hattori T., Uchiyama T., Yamanishi T.: Urinary function in the elderly with and without leukoaraiosis; in relation to cognitive and gait function. *J. Neurol. Neurosurg. Psychiatry*, 67: 5; 658-660 1999.

[35] Yamanishi T., Yasuda K., Sakakibara R., Suda S., Ishikawa N., Hattori T., Hosaka H.: Induction of urethral closure and inhibition of bladder contraction by continuous magnetic stimulation. *Neurourol. Urodynam.*, 18: 505-510 1999.

[36] Yamanishi T., Yasuda K., Sakakibara R., Hattori T., Minamide M., Yuki T., Ito H.: Variation in urinary flow according to voiding position in normal males. *Neurourol. Urodynam.*, :18: 553-557 1999.

[37] Sakakibara R., Hattori T., Mizobuchi K., Kuwabara S., Ogawa M.: Axonal poyneuropathy and encephalopathy in a patient with verotoxin-producing E. coli (VTEC) infection. *J. Neurol. Neurosurg. Psychiatry,* 67: 2; 254 1999.

[38] Sakakibara R., Fowler C. J., Hattori T.: Voiding and MRI analysis of the brain. *Int. Urogynaecol.*, J 10: 192-199 1999.

[39] Sakakibara R., Hattori T., Uchiyama T., Kita K., Asahina M., Suzuki A., Yamanishi T.: Urinary dysfunction and orthostatic hypotension in multiple system atrophy; which is the more common and earlier manifestation? *J. Neurol. Neurosurg. Psychiatry*, 68: 1; 65-69 2000.

[40] Sakakibara R., Hattori T., Uchiyama T., Yamanishi T.: Micturitional disturbance in pure autonomic failure. *Neurology*, 54: 499-501 2000.

[41] Sakakibara R., Hattori T., Uchiyama T., Suenaga T., Takahashi H., Yamanishi T., Egoshi K., Sekita N.: Are alpha-blockers involved in lower urinary tract dysfunction in multiple system atrophy? A comparison of prazosin and moxisylyte. *J. Auton. Nerv. Syst.*, 79: 191-195 2000.

[42] Sakakibara R., Fowler C. J., Hattori T., Hussain I. F., Swinn M. J., Uchiyama T., Yamanishi T.: Pressure-flow study as an evaluating method of neurogenic urethral relaxation failure. *J. Auton. Nerv. Syst.*, 80: 85-88 2000.

[43] Sakakibara R., Hattori T., Boku K., Uchiyama T., Yamanishi T.: Micturitional disturbance in neuro-Behcet's syndrome. *J. Auton. Nerv. Syst.*, 83: 86-89 2000.

[44] Yamanishi T., Yasuda K., Suda S., Ishikawa N., Sakakibara R., Hattori T.: Effect of functional continuous magnetic stimulation for urinary incontinence. *J. Urol.*, 163: 456-459 2000.

[45] Yamanishi T., Yasuda K., Sakakibara R., Hattori T., Tojo M.: The effectiveness of terazosin, an □1-blocker, on bladder neck obstruction as assessed by urodynamic hydraulic energy. *BJU International*, 85: 249-253 2000.

[46] Yamanishi T., Yasuda K., Sakakibara R., Hattori T., Suda S.: Randomized, double-blind study of electrical stimulation for urinary incontinence due to detrusor overactivity. *Urology*, 55: 353-357 2000.

[47] Yamanishi T., Sakakibara R., Uchiyama T., Suda S., Hattori T., Ito H., Yasuda K.: Comparative study of the effects of magnetic versus

electrical stimulation on inhibition of detrusor overactivity. *Urology*, 56: 777-781 2000.

[48] Yamanishi T., Yasuda K., Hamano S., Murayama N., Sakakibara R., Uchiyama T., Hattori T., Ito H.: Urethral obstruction in patients with nighttime wetting; urodynamic evaluation and outcome of surgical incision. *Neurourol. Urodynam.*, 19: 241-248 2000.

[49] Yamanishi T., Yasuda K., Murayama N., Sakakibara R., Uchiyama T., Ito H.: Biofeedback training for detrusor overactivity in children. *J. Urol.*, 164: 1686-1690 2000.

[50] Sakakibara R., Hattori T., Uchiyama T., Yamanishi T.: Micturitional disturbance in a patient with neurosarcoidosis. *Neurourol. Urodynam.*, 19: 273-277 2000.

[51] Sakakibara R., Hattori T., Uchiyama T., Yamanishi T., Ito H., Ito K.: Neurogenic failure of the external urethral sphincter closure and relaxation; a videourodynamic study Auton. *Neurosci. Basic Clin.*, 86: 208-215 2001.

[52] Sakakibara R., Hattori T., Uchiyama T., Yamanishi T.: Urinary dysfunction in Brown-Séquard syndrome. *Neurourol. Urodynam.*, 20: 661-667 2001.

[53] Sakakibara R., Hattori T., Uchiyama T., Yamanishi T.: Micturitional disturbance in subacute myelo-optico-neuropathy (SMON). *Auton. Neurosci. Basic Clin.*, 87: 282-285 2001.

[54] Aswal B. S., Berkley K. J., Hussain I., Brennan A., Craggs M., Sakakibara R., Frackowiak R. S. J., Fowler C. J.: Brain responses to changes in bladder volume and urge to void in healthy men. *Brain*, 124; 369-377 2001.

[55] Sakakibara R., Shinotoh H., Uchiyama T., Yoshiyama M., Hattori T., Yamanishi T.: SPECT imaging of the dopamine transporter with [123I]-□-CIT reveals marked decline of nigrostriatal dopaminergic function in Parkinson's disease with urinary dysfunction. *J. Neurol. Sci.*, 187: 55-59 2001.

[56] Sakakibara R., Hattori T., Uchiyama T., Yamanishi T.: Videourodynamic and sphincter motor unit potential analyses in Parkinson's disease and multiple system atrophy. *J. Neurol. Neurosurg. Psychiatry*, 71: 5; 600-606 2001 (with Editorial).

[57] Sakakibara R., Shinotoh H., Uchiyama T., Sakuma M., Kashiwado M., Yoshiyama M., Hattori T.: Questionnaire-based assessment of pelvic organ dysfunction in Parkinson's disease. *Auton. Neurosci. Basic Clin.*, 92; 76-85 2001.

[58] Sakakibara R., Uchiyama T., Kuwabara S., Kawaguchi N., Nemoto I., Nakata M., Hattori T.: Autonomic dysreflexia due to neurogenic bladder dysfunction; an unusual manifestation of spinal cord sarcoidosis. *J. Neurol. Neurosurg. Psychiatry,* 71: 6; 819-820 2001.

[59] Sakakibara R., Arai K., Fukutake T., Katayama K., Mori M., Hattori T.: Unilateral caudate head lesion simulating brain tumour in X-linked adult onset adrenoleukodystrophy. *J. Neurol. Neurosurg. Psychiatry,* 70: 3; 414-415 2001.

[60] Sakakibara R., Nakazawa K., Shiba K., Nakajima Y., Uchiyama T., Yoshiyama M., Yamanishi T., Hattori T.: Firing patterns of micturition-related neurons in the pontine storage centre in cats. *Auton. Neurosci. Basic Clin.,* 99: 24-30 2002.

[61] Sakakibara R., Nakazawa K., Uchiyama T., Yoshiyama M., Yamanishi T., Hattori T.: Micturition-related electrophysiological properties in the substantia nigra pars compacta and the ventral tegmental area in cats. *Auton. Neurosci. Basic Clin.,* 102: 30-38 2002.

[62] Sakakibara R., Odaka T., Uchiyama T., Asahina M., Yamaguchi K., Yamaguchi T., Yamanishi T., Hattori T.: Colonic transit time and rectoanal videomanometry in Parkinson's disease. *J. Neurol. Neurosurg. Psychiatry,* 74: 2; 268-272 2003.

[63] Sakakibara R., Uchiyama T., Asahina M., Yamanishi T., Hattori T.: Amezinium metilsulfate, a sympathomimetic agent, may increase the risk of urinary retention in multiple system atrophy. *Clin. Auton. Res.,* 13: 51-53 2003.

[64] Sakakibara R., Matsuda S., Uchiyama T., Yoshiyama M., Yamanishi T., Hattori T.: The effect of intranasal desmopressin on nocturnal waking in urination in multiple system atrophy patients with nocturnal polyuria. *Clin. Auton. Res.,* 13: 106-108 2003 (with Editorial).

[65] Sakakibara R., Hattori T., Uchiyama T., Kamura K., Yamanishi T.: Uro-neurological assessment in patients with spinal bifida occulta and cystica. *Neurourol. Urodynam.,* 22: 328-334 2003.

[66] Uchiyama T., Sakakibara R., Yamanishi T., Hattori T.: Short-term effect of l-dopa on the micturitional function in patients with Parkinson's disease. *Mov. Disord.,* 18: 573-578 2003.

[67] Sakakibara R., Hattori T., Uchiyama T., Yamanishi T.: Micturitional disturbance in patients with systemic lupus erythematosis. *Neurourol. Urodynam.,* 22: 593-596 2003.

[68] Sakakibara R., Nakazawa K., Uchiyama T., Yoshiyama M., Yamanishi T., Hattori T.: Effect of subthalamic nucleus stimulation on the Micturition reflex in cats. *Neuroscience,* 120: 871-875 2003.

[69] Hiraga A., Nagumo K., Sakakibara R., Kojima S., Fujinawa N., Hashimoto T.: Loss of urinary voiding sensation due to herpes zoster. *Neurourol. Urodynam.,* 22: 335-337 2003.

[70] Yamanishi T., Yasuda K., Yuki T., Sakakibara R., Uchiyama T., Kamai T., Tsuji T., Yoshida K.: Urodynamic evaluation of surgical outcome in patients with urinary retention due to central lumbar disc prolapse. *Neurourol. Urodynam.,* 22: 670-675 2003.

[71] Ito S., Kuwabara S., Sakakibara R., Ohki T., Arai H., Oda S., Hattori T.: Combined treatment with LDL-apheresis, chenodeoxycholic acid and HMG-CoA reductase inhibitor for cerebrotendinous xanthomatosis. *J. Neurol. Sci.,* 216: 179-182 2003.

[72] Sakakibara R., Hattori T., Uchiyama T., Yamanishi T.: Micturitional disturbance in a patient with a spinal cavernous angioma. *Neurourol. Urodynam.,* 22: 606-610 2003.

[73] Ito S., Sakakibara R., Yoshiyama Y., Hattori T.: Senile portoystemic hepatic encephalopathy as a treatable dementia-like syndrome. *J. Neurol.,* 251: 1015-1016 2004.

[74] Uchiyama T., Sakakibara R., Yamanishi T., Hattori T.: Lower urinary tract dysfunctions in patients with spinal cord tumors. *Neurourol. Urodynam.,* 23: 68-75 2004.

[75] Sakakibara R., Uchiyama T., Arai K., Yamanishi T., Hattori T.: Lower urinary tract dysfunction in Machado-Joseph disease: a study of 11 clinical-urodynamic observations. *J. Neurol. Sci.,* 218: 67-72 2004.

[76] Sakakibara R., Uchiyama T., Asahina M., Suzuki A., Yamanishi T., Hattori T.: Micturitional disturbance in acute idiopathic autonomic neuropathy. *J. Neurol. Neurosurg. Psychiatry,* 75: 287-291 2004.

[77] Sakakibara R., Uchiyama T., Yamanishi T., Hattori T.: Urinary function in patients with corticobasal degeneration: comparison with normal subjects. *Neurourol. Urodynam.,* 23: 154-158 2004.

[78] Sakakibara R., Hiruma K., Arai K., Uchiyama T., Hattori T.: Head-turning dizziness in multiple system atrophy. *Parkinson Dis. Related Disord.,* 10: 255-256 2004.

[79] Yamanishi T., Yasuda K., Kamai T., Tsuji T., Sakakibara R., Uchiyama T., Yoshida K.: Combination of a cholinergic drug and an alpha-blocker is more effective than monotherapy for the treatment of voiding

difficulty in patients with underactive detrusor. *Int. J. Urol.,* 11: 88-96 2004.

[80] Yamanishi T., Yasuda K., Kamai T., Tsuji T., Sakakibara R., Uchiyama T., Yoshida K. ﹕ Single-blind, randomized controlled study of the clinical and urodynamic effects of an alpha-blocker (naftopidil) and phytotherapy (eviprostat) in the treatment of benign prostatic hyperplasia. *Int. J. Urol.,* 11: 501-509 2004.

[81] Sakakibara R., Hirano S., Asahina M., Sawai S., Nemoto Y., Hiraga A., Uchiyama T., Hattori T.: Primary Sjogren's syndrome presenting with generalized autonomic failure. *Eur. J. Neurol.,* 11: 635-638 2004.

[82] Liu Z., Sakakibara R., Nakazawa K., Uchiyama T., Yamamoto T., Ito T., Hattori T.: Micturition-related neuronal firing in the periaqueductal gray area in cats. *Neuroscience,* 126: 1075-1082 2004.

[83] Sakakibara R., Odaka T., Uchiyama T., Asahina M., Yamaguchi K., Yamaguchi T., Yamanishi T., Hattori T.: Colonic transit time, sphincter EMG and rectoanal videomanometry in multiple system atrophy. *Mov. Disord.,* 19: 8; 924-929 2004.

[84] Sakakibara R., Uchida Y., Uchiyama T., Yoshiyama M., Yamanishi T., Hattori T.: Reduced cerebellar vermis activation in response to micturition in multiple system atrophy; 99mTc-labeled ECD SPECT study. *Eur. J. Neurol.,* 11: 705-708 2004.

[85] Ito T., Sakakibara R., Sakakibara Y., Mori M., Hattori T. ﹕ Medulla and gut. *Intern. Med.,* 43: 11; 1091 2004.

[86] Liu, Z., Sakakibara R., Odaka T., Uchiyama T., Yamamoto T., Ito T., Asahina M., Yamaguchi K., Yamaguchi T., Hattori T. ﹕ Mosapride citrate, a novel 5-HT4 agonist and partial 5-HT3 antagonist, ameliorates constipation in parkinsonian patients. *Mov. Disord.,* 20: 6; 680-686 2005.

[87] Sakakibara R., Odaka T., Liu Z., Uchiyama T., Yamamoto T., Ito T., Asahina M., Yamaguchi K., Yamaguchi T., Hattori T. ﹕ Dietary herb extract Dai-Kenchu-To ameliorates constipation in parkinsonian patients. *Mov. Disord.,* 20: 261-262 2005.

[88] Uchiyama T., Sakakibara R., Asahina M., Yamanishi T., Hattori T.: Post-micturitional hypotension in patients with multiple system atrophy. *J. Neurol. Neurosurg. Psychiatry,* 76: 186-190 2005.

[89] Sakakibara R., Ito T., Uchiyama T., Asahina M., Liu Z., Yamamoto T., Yamanaka Y., Hattori T.: Lower urinary tract function in dementia of

Lewy body type (DLB). *J. Neurol. Neurosurg. Psychiatry*, 76: 729-732 2005.

[90] Sakakibara R., Uchiyama T., Yoshiyama M., Yamanishi T., Hattori T.: Preliminary communication: urodynamic assessment of donepezil hydrochloride in patients with Alzheimer's disease. *Neurourol. Urodynam.*, 24: 273-275 2005.

[91] Yamamoto T., Sakakibara R., Uchiyama T., Liu Z., Ito T., Yamanishi T., Hattori T.: Lower urinary tract function in patients with pituitary adenoma compressing hypothalamus. *J. Neurol. Neurosurg. Psychiatry*, 76: 390-394 2005.

[92] Sakakibara R., Uchiyama T., Liu Z., Yamamoto T., Ito T., Yamanishi T., Hattori T.: Nocturnal polyuria with abnormal circadian rhythm of plasma arignin vasopressin in post-stroke patients. *Internal. Medicine*, 44: 4; 281-284 2005.

[93] Sakakibara R., Yamamoto T., Uchiyama T., Liu Z., Ito T., Yamazaki M., Awa Y., Yamanishi T., Hattori T.: Is lumbar spondylosis a cause of urinary retention in elderly women? *J. Neurol.*, 252: 953-957 2005.

[94] Sakakibara R., Hamano S., Uchiyama T., Liu Z., Yamanishi T., Hattori T.: Do BPH patients have neurogenic detrusor dysfunction? A uro-neurological assessment. *Urol. Int.*, 74: 44-50 2005.

[95] Yamamoto T., Sakakibara R., Hashimoto K., Nakazawa K., Uchiyama T., Liu Z., Ito T., Hattori T.: Striatal dopamine level increases in the urinary storage phase in cats: an in vivo microdialysis study. *Neuroscience*, 135: 299-303 2005.

[96] Yamamoto T., Sakakibara R., Uchiyama T., Liu Z., Ito T., Awa Y., Yamanishi T., Hattori T.: When is Onuf's nucleus involved in multiple system atrophy; a sphincter electromyography study. *J. Neurol. Neurosurg. Psychiatry*, 76: 1645-1648 2005.

[97] Sakakibara R., Uchiyama T., Liu Z., Yamamoto T., Ito T., Uzawa A., Suenaga T., Kanai K., Awa Y., Sugiyama Y., Hattori T.: Meningitis-retention syndrome; an unrecognized clinical condition. *J. Neurol.*, 252: 1495-1499 2005.

[98] Uzawa A., Sakakibara R., Tamura N., Asahina M., Yamanaka Y., Uchiyama T., Ito T., Yamamoto T., Liu Z., Hattori T.: Laryngeal abductor paralysis can be a solitary manifestation of multiple system atrophy. *J. Neurol. Neurosurg. Psychiatry*, 76: 1739-1741 2005.

[99] Hiraga A., Sakakibara R., Mori M., Suzuki A., Hattori T.: Bilateral lesion in the lateral columns and complete urinary retention: association

with the spinal cord descending pathway for micturition. *Neurourol. Urodynam.*, 24: 398-399 2005.

[100] Ogawa Y., Sakakibara R.: Platysma sign in high cervical lesion. *J. Neurol. Neurosurg. Psychiatry,* 76: 735 2005.

[101] Liu Z., Sakakibara R., Odaka T., Uchiyama T., Yamamoto T., Ito T., Hattori T.: Mechanism of abdominal massage for difficult defecation in a patient with myelopathy (HAM/TSP). *J. Neurol.,* 252: 1280-1282 2005.

[102] Sakakibara R., Ito T., Yamamoto T., Uchiyama T., Liu Z., Hattori T.: Vergence paresis in multiple system atrophy. *Intern. Med.,* 44: 911-912 2005.

[103] Ito T., Sakakibara R., Uchiyama T., Liu Z., Yamamoto T., Kashiwado K., Hattori T.: Lower urinary tract dysfunction in central pontine myelinolysis: possible contribution of the pontine micturition centre. *Eur. J. Neurol.,* 12: 812-813 2005.

[104] Hiraga A., Sakakibara R., Mizobuchi K., Asahina M., Kuwabara S., Hayashi Y., Hattori T.: Putaminal hemorrhage disrupts thalamocortical projection to secondary somatosensory cortex: case report. *J. Neurol. Sci.,* 231: 81-83 2005.

[105] Ito T., Sakakibara R., Uchiyama T., Liu Z., Yamamoto T., Hattori T.: Videomanometry of the pelvic organs; a comparison of the normal lower urinary and gastrointestinal tracts. *Int. J. Urol.,* 13: 29-35 2006.

[106] Ito T., Sakakibara R., Yamamoto T., Uchiyama T., Liu Z., Asahina M., Higashi M., Arai K., Ito S., Awa Y., Yamamoto K., Kinou M., Yamanishi T., Hattori T.: Urinary dysfunction and autonomic control in amyloid neuropathy. *Clin. Auton. Res.,* 16: 66-71 2006.

[107] Kanesaka T., Sakakibara R., Ito S., Ito T., Odaka T., Yamaguchi T., Uchiyama T., Liu Z., Yamamoto T., Hattori T.: Intestinal pseudo-obstruction in acute myelitis. *Intern. Med.,* 45: 35-36 2006.

[108] Liu Z., Sakakibara R., Uchiyama T., Yamamoto T., Ito T., Ito S., Awa Y., Odaka T., Yamaguchi T., Hattori T. Bowel dysfunction in Wolfram syndrome. *Diabetes Care,* 29: 2; 472-473 2006.

[109] Yamamoto T., Sakakibara R., Uchiyama T., Liu Z., Ito T., Awa Y., Yamanishi T., Hattori T.: Neurological diseases that cause detrusor hyperactivity with impaired contractile function. *Neurourol. Urodynam.,* 25: 4; 356-360 2006.

[110] Yamamoto T., Sakakibara R., Yamanaka Y., Uchiyama T., Asahina M., Liu Z., Ito T., Koyama Y., Awa Y., Yamamoto K., Kinou M., Hattori T.: Pyridostigmine in autonomic failure: can we treat postural hypotension

and bladder dysfunction with one drug? *Clin. Auton. Res.,* 16: 4; 296-298 2006.

[111] Sakakibara R., Yamaguchi T., Uchiyama T., Liu Z., Yamamoto T., Ito T., Okada T., Hattori T.: Calcium polycarbophil improves constipation in non-traumatic spinal cord disorders. *Clin. Auton. Res.,* 16: 4; 289-292 2006.

[112] Ito T., Sakakibara R., Yasuda K., Yamamoto T., Uchiyama T., Liu Z., Yamanishi T., Awa Y., Yamamoto K., Hattori T.: Incomplete emptying and urinary retention in multiple system atrophy: when does it occur and how do we manage it? *Mov. Disord.,* 21: 6; 816-823 2006.

[113] Yamanaka Y., Sakakibara R., Asahina M., Uchiyama T., Liu Z., Yamamoto T., Ito S., Suenaga T., Odata T., Yamaguchi T., Uehara K., Hattori T.: Chronic intestinal pseudo-obstruction as the initial feature of pure autonomic failure. *J. Neurol. Neurosurg. Psychiatry,* 77: 6; 800 2006.

[114] Sakakibara R., Yamaguchi C., Yamamoto T., Uchiyama T., Liu Z., Ito T., Awa Y., Yamamoto K., Kinou K., Hattori T.: Imidapril, an ACE inhibitor, could reverse loss of bladder sensation. *J. Neurol. Neurosurg. Psychiatry,* 77: 1100-1101 2006.

[115] Ito T., Sakakibara R., Ito S., Uchiyama T., Liu Z., Yamamoto T., Yamaguchi T., Odaka T., Higashi M., Hattori T. Mechanism of constipation in familial amyloid polyneuropathy: a case report. *Intern. Med.,* 45: 20; 1173-1175 2006.

[116] Asahina M., Hiraga A., Hayashi Y., Mizobuchi K., Sakakibara R., Lee K., Hattori T.: Ischemic electrocardiographic change induced by exercise in a patient with chronic autonomic failure. *Clin. Auton. Res.,* 16: 72-75 2006.

[117] Shimada J., Sakakibara R., Uchiyama T., Liu Z., Yamamoto T., Ito T., Mori M., Asahina M., Hattori T.: Intestinal pseudo-obstruction and neuroleptic malignant syndrome in a case of parkinsonian patient with chronic constipation. *Eur. J. Neurol.,* 13: 306-312 2006.

[118] Ito S., Sakakibara R., Komatsuzaki A., Nakata M., Uchiyama T., Hiruma K., Hattori T.: Aseptic meningoencephalitis presenting with bilateral vestibular ataxia: a case report. *Internal. Medicine,* 45: 8; 551-552 2006.

[119] Sakakibara R., Yamanishi T., Uchiyama T., Hattori T.: Acute urinary retention due to benign inflammatory nervous diseases. *J. Neurol.,* 253: 8; 1103-1110 2006.

[120] Sakakibara R., Uchiyama T., Tamura N., Kuwabara S., Asahina M., Hattori T. Urinary retention and sympathetic sphincter obstruction in axonal Guillain-Barré syndrome. *Muscle Nerve*, 35: 1; 111-115 2007.

[121] Yamaguchi C., Sakakibara R., Uchiyama T., Liu Z., Yamamoto T., Ito T., Awa Y., Yamamoto K., Kinou M., Yamanishi T., Nomura F., Hattori T. Bladder sensation in peripheral nerve lesions. *Neurourol. Urodyn.*, 25: 7; 763-769 2006.

[122] Ito T., Sakakibara R., Nakazawa K., Uchiyama T., Yamamoto T., Liu Z., Shimizu E., Hattori T. Effects of electrical stimulation of the raphe area on the micturition reflex in cats. *Neuroscience,* 142: 4; 1273-1280 2006.

[123] Sawai S., Sakakibara R., Kanai K., Kawaguchi N., Uchiyama T., Yamamoto T., Ito T., Liu Z., Hattori T. Isolated vomiting due to a unilateral dorsal vagal complex lesion. *Eur. Neurol.,* 56: 4; 246-248 2006.

[124] Hiraga A., Sakakibara R., Mori M., Yamanaka Y., Ito S., Hattori T. Urinary retention can be the sole initial manifestation of acute myelitis. *J. Neurol. Sci.,* 251: 1-2; 110-112 2006.

[125] Sakakibara R., Yamaguchi T., Uchiyama T., Yamamoto T., Ito T., Liu Z., Odaka T., Yamaguchi C., Hattori T. Calcium polycarbophil improves constipation in primary autonomic failure and multiple system atrophy subjects. *Mov. Disord.,* 22: 11; 1672-1673 2007.

[126] Kanai K., Sakakibara R., Uchiyama T., Liu Z., Yamamoto T., Ito T., Hirano S., Asahina M., Kuwabara S., Hattori T., Fukami G., Arai K., Yamaguchi C., Nomura F. Sporadic case of spinocerebellar ataxia type 17: treatment observations for managing urinary and psychotic symptoms. *Mov. Disord.,* 22: 3; 441-443 2007.

[127] Yamanaka Y., Asahina M., Hiraga A., Sakakibara R., Oka H., Hattori T. Over 10 years of isolated autonomic failure preceding dementia and Parkinsonism in 2 patients with Lewy body disease. *Mov. Disord.,* 22: 4; 595-597 2007.

[128] Asahina M., Sakakibara R., Liu Z., Ito T., Yamanaka Y., Nakazawa K., Shimizu E., Hattori T. The raphe magnus/pallidus regulates sweat secretion and skin vasodilation of the cat forepaw pad: a preliminary electrical stimulation study. *Neurosci. Lett.,* 415: 3; 283-287 2007.

[129] Ito S., Sakakibara R., Hattori T. Wolfram syndrome presenting marked brain MR imaging abnormalities with few neurologic abnormalities. *AJNR Am. J. Neuroradiol.,* 28: 2; 305-306 2007.

[130] Sawai S., Sakakibara R., Uchiyama T., Liu Z., Yamamoto T., Ito T., Kuwabara S., Kanai K., Asahina M., Yamanaka T., Odaka T.,

Yamaguchi T., Hattori T. Acute motor axonal neuropathy presenting with bowel, bladder, and erectile dysfunction. *J. Neurol.*, 254: 2; 250-252 2007.

[131] Yamaguchi C., Sakakibara R., Uchiyama T., Yamamoto T., Ito T., Liu Z., Awa Y., Yamamoto K., Nomura F., Yamanishi T., Hattori T. Overactive bladder in diabetes: a peripheral or central mechanism? *Neurourol. Urodyn.,* 26: 6; 807-813 2007.

[132] Sakakibara R., Uchiyama T., Awa Y., Liu Z., Yamamoto T., Ito T., Yamamoto K., Kinou M., Yamaguchi C., Yamanishi T., Hattori T. Psychogenic urinary dysfunction: a uro-neurological assessment. *Neurourol. Urodyn.*, 26: 4; 518-524 2007.

[133] Sakakibara R., Uchiyama T., Yamaguchi C., Yamamoto T., Ito T., Liu Z., Awa Y., Yamazaki M., Hattori T. Urinary retention due to an isolated sacral root injury caused by sacral fracture. *Spinal Cord,* 45: 12; 790-792 2007.

[134] Sakakibara R., Murakami E., Katagiri A., Hayakawa S., Uchiyama T., Yamamoto T., Hattori T. Moxibustion, an alternative therapy, ameliorated disturbed circadian rhythm of plasma arginine vasopressin and urine output in multiple system atrophy. *Intern. Med.*, 46: 13; 1015-1018 2007.

[135] Mashidori T., Yamanishi T., Yoshida K., Sakakibara R., Sakurai K., Hirata K. Continuous urinary incontinence presenting as the initial symptoms demonstrating acontractile detrusor and intrinsic sphincter deficiency in multiple system atrophy. *Int. J. Urol.,* 14: 10; 972-974 2007.

[136] Sakakibara R., Awa Y., Naya Y., Tobe T., Uchiyama T., Hattori T. Neobladder overactivity; an equivalent to spontaneous rectal contraction. *Int. J. Urol.*, 14: 11; 1054-1056 2007.

[137] Yamanishi T., Kamai T., Sakakibara R., Uchiyama T., Yamamoto T., Ito T., Yoshida K. Which □-adrenoceptor subtypes are important in the treatment of overactive bladder? *Current Drug Therapy*, 2: 79-84 2007.

[138] Sakakibara R., Ito T., Uchiyama T., Awa Y., Yamaguchi C., Hattori T. Effects of milnacipran and paroxetine on overactive bladder due to neurologic diseases: a urodynamic assessment. *Urol. Int.,* 2008; 81(3): 335-9.

[139] Tateno F., Sakakibara R., Saiki A., Miyashita Y., Shirai K. Levodopa might affect metaiodobenzylguanidine myocardial accumulation. *Mov. Disord.*, 2008 Oct. 30;23(14):2097-8.

[140] Sakakibara R., Uchiyama T., Yamanishi T., Kishi M. Dementia and lower urinary dysfunction: with a reference to anticholinergic use in elderly population. *Int. J. Urol.*, 2008 Sep.; 15(9):778-88.

[141] Sakakibara R., Yamaguchi C., Uchiyama T., Ito T., Liu Z., Yamamoto T., Awa Y., Yamanishi T., Hattori T. Pelvic autonomic dysfunction without paraplegia: a sequel of spinal cord stroke. *Eur. Neurol.*, 2008; 60(2):97-100.

[142] Oyama T., Shirai K., Sakakibara R., Kishi M., Ogawa E., Suzuki Y. Multiple system atrophy and impaired gallbladder emptying. *Mov. Disord.*, 2008 Jul. 15; 23(9):1321-2.

[143] Endo K., Shirai K., Sakakibara R. Multiple system atrophy and the syndrome of inappropriate secretion of antidiuretic hormone. *Mov Disord.*, 2008 Jul. 15; 23(9):1325-6.

[144] Sakakibara R., Yamazaki M., Mannouji C., Yamaguchi C., Uchiyama T., Ito T., Liu Z., Yamamoto T., Awa Y., Yamanishi T., Hattori T. Urinary retention without tetraparesis as a sequel to spontaneous spinal epidural hematoma. *Intern. Med.*, 2008;47(7):655-7.

[145] Sakakibara R., Uchiyama T., Yamanishi T., Shirai K., Hattori T. Bladder and bowel dysfunction in Parkinson's disease. *J. Neural. Transm.*, 2008; 115(3):443-60.

[146] Sakakibara R., Kanda T., Sekido T., Uchiyama T., Awa Y., Ito T., Liu Z., Yamamoto T., Yamanishi T., Yuasa T., Shirai K., Hattori T. Mechanism of bladder dysfunction in idiopathic normal pressure hydrocephalus. *Neurourol. Urodyn.*, 2008; 27(6):507-10.

[147] Sakakibara R., Ito T., Uchiyama T., Awa Y., Yamaguchi C., Hattori T. Effects of milnacipran and paroxetine on overactive bladder due to neurologic diseases: a urodynamic assessment. *Urol. Int.*, 2008; 81(3): 335-9.

[148] Yamanishi T., Tatsumiya K., Furuya N., Masuda A., Kamai T., Sakakibara R., Uchiyama T., Yoshida K. Long-term efficacy of tamsulosin in the treatment of lower urinary tract symptoms suggestive of benign prostate hyperplasia in real-life practice. *Uro. Today International*, 2009: in press. Doi 10.3834.

[149] Yamanishi T., Masuda A., Mizuno T., Kamai T., Tatsumiya K., Fukuda T., Furuya N., Watanabe M., Sakakibara R., Uchiyama T., Yoshida K. Amburatory urodynamics in asymptomatic, young, healthy male volunteers. *LUTS*, 2009; 1: 29-34.

[150] Sakakibara R., Ogata T., Uchiyama T., Kishi M., Ogawa E., Isaka S., Yuasa J., Yamamoto T., Ito T., Yamanishi T., Awa Y., Yamaguchi C.,

Takahashi O. How to manage overactive bladder in elderly individuals with dementia? A combined use of donepezil, a central AChE inhibitor, and propiverine, a peripheral muscarine receptor antagonist. *J. Am. Geriatr. Soc.,* 2009 Aug.; 57(8):1515-7.

[151] Yamamoto T., Sakakibara R., Uchiyama T., Liu Z., Ito T., Awa Y., Yamanishi T., Hattori T. Questionnaire-based assessment of pelvic organ dysfunction in multiple system atrophy. *Mov. Disord.,* 2009 May 15; 24(7):972-8.

[152] Yamamoto T., Sakakibara R., Nakazawa K., Uchiyama T., Shimizu E., Hattori T. Effects ofelectrical stimulation of the striatum on bladder activity in cats. *Neurourol. Urodyn.,* 2009; 28(6):549-54.

[153] Sakakibara R., Uchiyama T., Kuwabara S., Mori M., Ito T., Yamamoto T., Awa Y., Yamaguchi C., Yuki N., Vernino S., Kishi M., Shirai K. Prevalence and mechanism of bladder dysfunction in Guillain-Barré Syndrome. *Neurourol. Urodyn.,* 2009; 28(5):432-7.

[154] Furukawa R., Sakakibara R., Hosoe N., Kishi M., Ogawa E., Suzuki Y. Ataxia and middle cerebellar peduncle lesions in hepatic encephalopathy. *Neuroradiology,* 2009 Apr.; 51(4):273-4.

[155] Sakakibara R., Uchiyama T., Yamanishi T., Kishi M. Genito-urinary dysfunction in Parkinson's disease. *Clin. Auton. Res.,* 2009: in press.

[156] Sakakibara R., Kishi M., Ogawa E., Shirai K. Isolated facio-lingual hypoalgesia and weakness after a hemorrhagic infarct localized at the contralateral operculum. *J. Neurol. Sci.,* 2009 Jan. 15; 276(1-2):193-5.

[157] Sakakibara R., Uchiyama T., Yamanishi T., Kishi M. Sphincter EMG as a diagnostic tool in autonomic disorders. *Clin. Auton. Res.,* 2009 Feb.; 19(1):20-31.

[158] Matsuzawa Y., Sakakibara R., Shoda T., Kishi M., Ogawa E. Good maternal and fetal outcomes of predominantly sensory Guillain-Barré syndrome in pregnancy after intravenous immunoglobulin. *Neurol. Sci.,* 2010 Apr.; 31(2):201-3.

[159] Uchiyama T., Sakakibara R., Yamamoto T., Ito T., M. CY, Awa Y., Yano M., Yanagisawa M., Kobayashi M., Higuchi Y., Ichikawa T., Yamanishi T., Hattori T., Kuwabara S. Comparing bromocriptine effects with levodopa effects on bladder function in Parkinson's disease. *Mov. Disord.,* 2009 Dec. 15; 24(16):2386-90.

[160] Hirano S., Asahina M., Uchida Y., Shimada H., Sakakibara R., Shinotoh H., Hattori T. Reduced perfusion in the anterior cingulate cortex of patients with pure autonomic failure: an 123I-IMP SPECT study. *J. Neurol. Neurosurg. Psychiatry,* 2009 Sep.; 80(9):1053-5.

[161] Sakakibara R., Koide N., Kishi M., Ogawa E., Shirai K. Aseptic meningitis as the sole manifestation of Behçet's disease. *Neurol. Sci.,* 2009 Oct.; 30(5):405-7.

[162] Ogawa E., Sakakibara R., Kishi M., Shirai K. Exercise-induced hypertension in pure autonomic failure. *Eur. J. Neurol.,* 2009 Aug.; 16(8):e151-2.

[163] Uchiyama T., Sakakibara R., Yoshiyama M., Yamamoto T., Ito T., Liu Z., Yamaguchi C., Awa Y., Yano H. M., Yanagisawa M., Yamanishi T., Hattori T., Kuwabara S. Biphasic effect of apomorphine, an anti-parkinsonian drug, on bladder function in rats. *Neuroscience,* 2009 Sep. 15; 162(4):1333-8.

[164] Homma Y., Yamaguchi O.; Imidafenacin Study Group. A randomized, double-blind, placebo- and propiverine-controlled trial of the novel antimuscarinic agent imidafenacin in Japanese patients with overactive bladder. *Int. J. Urol.,* 2009 May; 16(5):499-506.

[165] Kishi M., Sakakibara R., Nagao T., Terada H., Ogawa E. Thalamic infarction disrupts spinothalamocortical projection to the mid-cingulate cortex and supplementary motor area. *J. Neurol. Sci.,* 2009 Jun. 15; 281 (1-2):104-7.

[166] Takahashi O., Sakakibara R., Kishi M., Ogawa E., Tsunoyama K., Sugiyama M., Tomaru T., Uchiyama T., Yamanishi T. Pelvic autonomic dysfunction without tetraparesis: A sequel of rubella-related acute longitudinal myelitis. *LUTS,* 2009, 1; 103-106.

[167] Hosoe N., Sakakibara R., Yoshida M., Wakabayashi T., Kikuchi H., Yamada T., Kawahara K., Kishi M. Acute, severe constipation in a 58-year-old Japanese patient. *Gut,* 2010 Oct. 27. [Epub ahead of print] No abstract available. PMID: 20980341 [PubMed- as supplied by publisher.

[168] Watanabe M., Yamanishi T., Honda M., Sakakibara R., Uchiyama T., Yoshida K. Efficacy of extended-release tolterodine for the treatment of neurogenic detrusor overactivity and/or low-compliance bladder. *Int. J. Urol.,* 2010 Nov.; 17(11):931-6.

[169] Asahina M., Akaogi Y., Misawa S., Kanai K., Ando Y., Sakakibara R., Arai K., Hattori T., Kuwabara S. Sensorimotor manifestations without autonomic symptoms in two siblings with TTR Val107 familial amyloid polyneuropathy. *Clin. Neurol. Neurosurg.,* 2010 Oct. 11. [Epub ahead of print] No abstract available. PMID: 20943310 [PubMed - as supplied by publisher.

[170] Sakakibara R., Kishi M., Ogawa E., Tateno F., Ogata T., Haruta H., Matsuzawa Y., Shirai K. Dehydration encephalopathy: a neurological

emergency in the older adults. *J. Am. Geriatr. Soc.,* 2010 Sep.; 58(9): 1819-21.

[171] Takahashi O., Sakakibara R., Kishi M., Matsuzawa Y., Ogawa E., Sugiyama M., Uchiyama T., Yamamoto T., Yamanishi T., Tomaru T. Herbal medicine-induced meningitis-retention syndrome. *Intern. Med.,* 2010; 49(16):1813-6.

[172] Sakakibara R., Kishi M., Ogawa E., Yuasa J., Isaka S., Uchiyama T., Yamanishi T. Dentatorubral pallidoluysian atrophy presenting with urinary retention. *Mov. Disord.,* 2010 Sep. 15; 25(12):1996-7.

[173] Yamamoto T., Sakakibara R., Uchiyama T., Yamaguchi C., Nomura F., Ito T., Yanagisawa M., Yano M., Awa Y., Yamanishi T., Hattori T., Kuwabara S. Pelvic organ dysfunction is more prevalent and severe in MSA-P compared to parkinson's disease. *Neurourol. Urodyn.,* 2010 Jul. 23. [Epub ahead of print]PMID: 20658542 [PubMed - as supplied by publisher.

[174] Sakakibara R., Kishi M., Ogawa E., Takahashi O., Sugiyama M., Uchiyama T., Yamamoto T., Yamanishi T. Multiple-system atrophy presenting with low rectal compliance and bowel pain. *Mov. Disord.,* 2010 Jul. 30; 25(10):1516-8.

[175] Sakakibara R. Editorial comment to effect of dominant hemispheric stroke on detrusor function in patients with lower urinary tract symptoms. *Int. J. Urol.,* 2010 Jul.; 17(7):660.

[176] Sakakibara R., Tsunoyama K., Takahashi O., Sugiyama M., Kishi M., Ogawa E., Uchiyama T., Yamamoto T., Yamanishi T., Awa Y., Yamaguchi C. Real-time measurement of oxyhemoglobin concentration changes in the frontal micturition area: an fNIRS study. *Neurourol. Urodyn.,* 2010 Jun.; 29(5):757-64.

[177] Committee for Establishment of the Clinical Guidelines for Nocturia of the Neurogenic Bladder Society. Clinical guidelines for nocturia. *Int. J. Urol.,* 2010 May; 17(5):397-409.

[178] Tateno F., Sakakibara R., Kishi M., Ogawa E. Bupivacaine-induced chemical meningitis. *J. Neurol.,* 2010 Aug.; 257(8):1327-9.

[179] Yamamoto T., Sakakibara R., Nakazawa K., Uchiyama T., Shimizu E., Hattori T., Kuwabara S. Neuronal activities of forebrain structures with respect to bladder contraction in cats. *Neurosci. Lett.,* 2010 Mar. 31; 473(1):42-7.

[180] Sakakibara R., Uchiyama T., Yamanishi T., Kishi M. Genitourinary dysfunction in Parkinson's disease. *Mov. Disord.,* 2010 Jan. 15; 25(1): 2-12.

[181] Wyndaele J. J., Kovindha A., Igawa Y., Madersbacher H., Radziszewski P., Ruffion A., Schurch B., Castro D., Sakakibara R., Wein A.; ICI 2009 Committee 10. Neurologic fecal incontinence. *Neurourol. Urodyn.,* 2010; 29(1):207-12.

[182] Wyndaele J. J., Kovindha A., Madersbacher H., Radziszewski P., Ruffion A., Schurch B., Castro D., Igawa Y., Sakakibara R., Wein A. Committe 10 on Neurogenic Bladder and Bowel of the International Consultation on Incontinence 2008-2009. Neurologic urinary incontinence. *Neurourol. Urodyn.,* 2010; 29(1):159-64. Review.

[183] Matsuzawa Y., Sakakibara R., Shoda T., Kishi M., Ogawa E. Good maternal and fetal outcomes of predominantly sensory Guillain-Barré syndrome in pregnancy after intravenous immunoglobulin. *Neurol. Sci.,* 2010 Apr.; 31(2):201-3.

[184] Tateno F., Sakakibara R., Kishi M., Ogawa E., Yoshimatsu Y., Takada N., Suzuki Y., Mouri T., Uchiyama T., Yamamoto T. Incidence of emergency intestinal pseudo-obstruction in Parkinson's disease. *J. Am. Geriatr. Soc.,* 2011 Dec.; 59(12):2373-5.

[185] Sakakibara R., Uchida Y., Ishii K., Kazui H., Hashimoto M., Ishikawa M., Yuasa T., Kishi M., Ogawa E., Tateno F., Uchiyama T., Yamamoto T., Yamanishi T., Terada H.; the members of SINPHONI (Study of Idiopathic Normal Pressure Hydrocephalus On Neurological Improvement). Correlation of right frontal hypoperfusion and urinary dysfunction in iNPH: A SPECT study. *Neurourol. Urodyn.,* 2011 Oct 28. doi: 10.1002/nau.21222. [Epub ahead of print].

[186] Tateno F., Sakakibara R., Kawai T., Kishi M., Murano T. Alpha-synuclein in the Cerebrospinal Fluid Differentiates Synucleinopathies (Parkinson Disease, Dementia With Lewy Bodies, Multiple System Atrophy) From Alzheimer Disease. *Alzheimer Dis. Assoc. Disord.,* 2011 Oct 26. [Epub ahead of print].

[187] Sakakibara R., Tateno F., Kishi M., Tsuyuzaki Y., Uchiyama T., Yamamoto T. Pathophysiology of bladder dysfunction in Parkinson's disease. *Neurobiol. Dis.,* 2012 Jun.; 46(3):565-71.

[188] Ogawa E., Sakakibara R., Yoshimatsu Y., Suzuki Y., Mouri T., Tateno F., Kishi M., Oda S., Imamura H. Crohn's disease and stroke in a young adult. *Intern. Med.,* 2011; 50(20):2407-8.

[189] Sakakibara R., Kishi M., Ogawa E., Tateno F., Uchiyama T., Yamamoto T., Yamanishi T. Bladder, bowel, and sexual dysfunction in Parkinson's disease. *Parkinsons Dis.,* 2011; 2011:924605. Epub 2011 Sep. 12.

[190] Suzuki J., Sakakibara R., Tomaru T., Tateno F., Kishi M., Ogawa E., Kurosu T., Shirai K. Stroke and Cardio-ankle Vascular Stiffness Index. *J. Stroke Cerebrovasc. Dis.*, 2011 Aug. 19. [Epub ahead of print].

[191] Ogawa E., Sakakibara R., Kishi M., Tateno F. Pure isolated internuclear ophthalmoplegia. *Intern. Med.*, 2011; 50(16):1785.

[192] Kishi M., Sakakibara R., Ogawa E., Tateno F., Takahashi O., Koga M. Bilateral abducens palsy in a case of cytomegalovirus-associated Guillain-Barré syndrome. *Neurol. Sci.*, 2011 Jul. 22. [Epub ahead of print].

[193] Ishii K., Hashimoto M., Hayashida K., Hashikawa K., Chang C. C., Nakagawara J., Nakayama T., Mori S., Sakakibara R. A Multicenter Brain Perfusion SPECT Study Evaluating Idiopathic Normal-Pressure Hydrocephalus on Neurological Improvement. *Dement. Geriatr. Cogn. Disord.*, 2011; 32(1):1-10. Epub 2011 Jul. 29.

[194] Ogawa E., Sakakibara R., Kishi M., Tateno F. Constipation triggered the malignant syndrome in Parkinson's disease. *Neurol. Sci.*, 2011 Jul. 20. [Epub ahead of print].

[195] Tateno F., Sakakibara R., Yokoi Y., Kishi M., Ogawa E., Uchiyama T., Yamamoto T., Yamanishi T., Takahashi O. Levodopa ameliorated anorectal constipation in de novo Parkinson's disease: The QL-GAT study. *Parkinsonism Relat. Disord.*, 2011 Jun. 24. [Epub ahead of print] PMID:21705259 [PubMed - as supplied by publisher].

[196] Tateno F., Sakakibara R., Kishi M., Ogawa E. Hashimoto's Ophthalmopathy. *Am. J. Med. Sci.*, 2011 Jul.; 342(1):83-5.

[197] Ogawa E., Sakakibara R., Kawashima K., Yoshida T., Kishi M., Tateno F., Kataoka M., Kawashima T., Yamamoto M. VGCC antibody-positive paraneoplastic cerebellar degeneration presenting with positioning vertigo. *Neurol. Sci.*, 2011 Jun. 16. [Epub ahead of print] PMID:21678073 [PubMed - as supplied by publisher].

[198] Tateno F., Sakakibara R., Sugiyama M., Takahashi O., Kishi M., Ogawa E., Uchiyama T., Yamamoto T., Yamanishi T., Yano H., Suzuki H. Meningitis-retention Syndrome: First Case of Urodynamic Follow-up. *Intern. Med.*, 2011; 50(12):1329-32.

[199] Uchiyama T., Sakakibara R., Yamamoto T., Ito T., Yamaguchi C., Awa Y., Yanagisawa M., Higuchi Y., Sato Y., Ichikawa T., Yamanishi T., Hattori T., Kuwabara S. Urinary dysfunction in early and untreated Parkinson's disease. *J. Neurol. Neurosurg. Psychiatry*, 2011 Jun. 13. [Epub ahead of print] PMID:21670077[PubMed - as supplied by publisher].

[200] Tateno F., Sakakibara R., Kishi M., Ogawa E. Pure akinesia with low myocardial metaiodobenzylguanidine uptake. *Parkinsonism Relat. Disord.*, 2011 Jun.; 17(5):357-9.

[201] Tateno F., Sakakibara R., Kishi M., Ogawa E., Terada H., Ogata T., Haruta H. Sensitivity and specificity of metaiodobenzylguanidine (MIBG) myocardial accumulation in the diagnosis of Lewy body diseases in a movement disorder clinic. *Parkinsonism Relat. Disord.*, 2011 Jun.; 17(5):395-7.

[202] Tsunoyama K., Sakakibara R., Yamaguchi C., Uchiyama T., Yamamoto T., Yamanishi T., Takahashi O., Sugiyama M., Kishi M., Ogawa E. Pathogenesis of reduced or increased bladder sensation. *Neurourol. Urodyn.*, 2011 Mar.; 30(3):339-43.

[203] Kishi M., Sakakibara R., Terada H., Ogawa E., Tateno T. Does levodopa affect metaiodobenzylguanidine myocardial accumulation in Parkinson's disease? *Mov. Disord.*, 2011 Feb. 15; 26(3):564-5.

[204] Tateno F., Sakakibara R., Kishi M., Yuasa T., Ogawa E., Takahashi O., Yoshio S., Sugiyama M., Uchiyama T., Yamamoto T., Yamanishi T. Progressive supranuclear palsy presenting with urinary retention and sleep apnea. *Clin. Auton. Res.*, 2011 Jan. 6. [Epub ahead of print]PMID:21210294 [PubMed - as supplied by publisher].

[205] Hosoe N., Sakakibara R., Yoshida M., Wakabayashi T., Kikuchi H., Yamada T., Kawahara K., Kishi M. Acute, severe constipation in a 58-year-old Japanese patient. *Gut,* 2011 Aug.; 60(8):1059, 1093.

[206] Ogawa E., Sakakibara R., Tateno F., Kishi M., Nakagami T., Terada H. A case of brainstem hypertensive encephalopathy. *Am. J. Case Reports,* 2011; 12: 95-97.

[207] Tateno F., Sakakibara R., Kishi M., Ogawa E., Tsuyusaki Y., Yoshida T., Yamamoto M. Zoster lower cranial neuropathy: a case report and a review of literature. *Am. J. Case Reports*, 2011; 12: 137-139.

[208] Tateno F., Sakakibara R., Kishi M., Ogawa E., Yoshimatsu Y., Takada N., Suzuki Y., Mouri T., Uchiyama T., Yamamoto T. Incidence of emergency intestinal pseudo-obstruction in Parkinson's disease. *J. Am. Geriatr. Soc.,* 2011 Dec.; 59(12):2373-5.

[209] Ogawa E., Sakakibara R., Endo K., Tateno F., Matsuzawa Y., Hosoe N., Kishi M., Shirai K. Incidence of dehydration encephalopathy among patients with disturbed consciousness at a hospital emergency unit. *Am. Clinics and Practice,* 2011; 1: e9.

[210] Kishi M., Sakakibara R., Nomura T., Yoshida T., Yamamoto M., Kataoka M., Ogawa E., Tateno F. Lateral medullary infarction

presenting as isolated vertigo and unilateral loss of visual suppression. *Neurol. Sci.,* (2012) 33:129–132.

[211] Takahashi O., Sakakibara R., Panicker J., Fowler C. J., Tateno F., Kishi M., Tsuyusaki Y., Yano H., Sugiyama M., Uchiyama T., Yamamoto T. White matter lesions or Alzheimer's disease: which contributes more to overactive bladder and incontinence in elderly adults with dementia? *J. Am. Geriatr. Soc.,* 2012; 60: 2370-2371.

[212] Sakakibara R., Panicker J., Fowler C. J., Tateno F., Kishi M., Tsuyuzaki Y., Ogawa E., Uchiyama T., Yamamoto T. Vascular incontinence: incontinence in the elderly due to ischemic white matter changes. *Neurol. Int.,* 2012 Jun. 14; 4(2):e13. doi: 10.4081/ni.2012.e13. Epub 2012 Sep 6.

[213] Terayama K., Sakakibara R., Ogawa A., Haruta H., Akiba T., Nagao T., Takahashi O., Sugiyama M., Tateno A., Tateno F., Yano M., Kishi M., Tsuyusaki Y., Uchiyama T., Yamamoto T. Weak detrusor contractility correlates with motor disorders in Parkinson's disease. *Mov. Disord.,* 2012 Oct. 18. doi: 10.1002/mds.25225. [Epub ahead of print].

[214] Doi H., Sakakibara R., Sato M., Masaka T., Kishi M., Tateno A., Tateno F., Tsuyusaki Y., Takahashi O. Plasma levodopa peak delay and impaired gastric emptying in Parkinson's disease. *J. Neurol. Sci.,* 2012; 319: 86-88.

[215] Yamamoto T., Sakakibara R., Uchiyama T., Yamaguchi C., Nomura F., Ito T., Yanagisawa M., Yano M., Awa Y., Yamanishi T., Hattori T., Kuwabara S. Receiver operating characteristic analysis of sphincter electromyography for parkinsonian syndrome. *Neurourol. Urodyn.,* 2012; 31: 1128-1134.

[216] Tateno F., Sakakibara R., Kishi M., Ogawa E., Takada N., Hosoe N., Suzuki Y., Takahashi M., Uchiyama T., Yamamoto T. Constipation and metaiodobenzylguanidine myocardial scintigraphy abnormality. *J. Am. Geriatr. Soc.,* 2012; 60: 185-187.

[217] Tateno F., Sakakibara R., Kishi M., Ogawa E., Yoshimatsu Y., Takada N., Suzuki Y., Mouri T., Uchiyama T., Yamamoto T. Incidence of emergency intestinal pseudo-obstruction in Parkinson's disease. *J. Am. Geriatr. Soc.,* 2011; 59: 2373-2375.

[218] Sakakibara R., Uchida Y., Ishii K., Kazui H., Hashimoto M., Ishikawa M., Yuasa T., Kishi M., Ogawa E., Tateno F., Uchiyama T., Yamamoto T., Yamanishi T., Terada H.; SINPHONI (Study of Idiopathic Normal Pressure Hydrocephalus On Neurological Improvement). Correlation of

right frontal hypoperfusion and urinary dysfunction in iNPH: a SPECT study. *Neurourol. Urodyn.*, 2012; 31: 50-55.

[219] Tateno F., Sakakibara R., Kawai T., Kishi M., Murano T. Alpha-synuclein in the cerebrospinal fluid differentiates synucleinopathies (Parkinson Disease, dementia with Lewy bodies, multiple system atrophy) from Alzheimer disease. *Alzheimer Dis. Assoc. Disord.*, 2012; 26: 213-216.

[220] Sakakibara R., Tateno F., Kishi M., Tsuyuzaki Y., Uchiyama T., Yamamoto T. Pathophysiology of bladder dysfunction in Parkinson's disease. *Neurobiol. Dis.*, 2012; 46: 565-571.

[221] Tateno F., Sakakibara R., Sugiyama M., Kishi K., Ogawa E., Takahashi O., Yano M., Uchiyama T., Yamamoto T., Tsuyuzaki Y. Lower urinary tract function in spinocerebellar ataxia 6. *LUTS,* 2012; 4: 41-44.

[222] Takahashi O., Sakakibara R., Tsunoyama K., Tateno F., Yano M., Sugiyama M., Uchiyama T., Yamamoto T., Awa Y., Yamaguchi C., Yamanishi T., Kishi K., Tsuyusaki Y. Do sacral/ peripheral lesions contribute to detrusor- sphincter dyssynergia? *LUTS*, 2012; 4: 126-129.

[223] Tateno F., Sakakibara R., Kishi K., Tsuyusaki Y., Furukawa R., Yoshimatsu Y., Suzuki Y. Brainstem stroke and increased anal tone. *LUTS*, 2012; 4: 161-163.

[224] Ito T., Sakakibara R., Shimizu E., Kishi K., Tsuyusaki Y., Tateno F., Uchiyama T., Yamamoto T. Is major depression a risk for bladder, bowel, and sexual dysfunction? *LUTS*, 2012; 4: 87-95.

[225] Sakakibara R., Tateno F., Tsuyusaki Y., Kishi M., Uchiyama T., Yamamoto T., Yamanishi T. Psychogenic urinary dysfunction in children and adults. *Curr. Bladder Dysfunct. Rep.*, 2012; 7: 242-246.

[226] Kishi M., Sakakibara R., Yoshida T., Yamamoto M., Suzuki M., Kataoka M., Tsuyusaki Y., Tateno A., Tateno F. Visual suppression is impaired in spinocerebellar ataxia type 6 but preserved in benign paroxysmal positional vertigo. *Diagnostics,* 2012; 2: 52-56.

[227] Sakakibara R., Panicker J., Fowler C. J., Tateno F., Kishi M., Tsuyusaki Y., Uchiyama T., Yamamoto T. "Vascular incontinence" and normal-pressure hydrocephalus: two common sources of elderly incontinence with brain etiologies. *Current Drug Therapy*, 2012; 7: 67-76.

[228] Kishi M., Sakakibara R., Inaoka T., Terada H., Rikitake H. Diffusion tensor imaging of peripheral nerves in Churg-Strauss syndrome. *Int. J. Sci.*, 2012: Dec. page 1-7. E-pub ahead of print.

[229] Sakakibara R., Ogata T., Haruta M., Kishi M., Tsuyusaki Y., Tateno A., Tateno F., Mouri T. Amnestic mild cognitive impairment with low

myocardial metaiodobenzylguanidine uptake. *Am. J. Neurodegener. Dis.*, 2012;1(2):146-151.

[230] Doi H., Sakakibara R., Sato M., Hirai S., Masaka T., Kishi M., Tsuyusaki Y., Tateno A., Tateno F., Takahashi O., Ogata T. Nizatidine ameliorates gastroparesis in Parkinson's disease: A pilot study. *Mov. Disord.*, 2013 Dec 27. doi: 10.1002/mds.25777. [Epub ahead of print]PMID:24375669[PubMed - as supplied by publisher].

[231] Sakakibara R., Tateno F., Kishi M., Tsuyusaki Y., Terada H., Inaoka T. MIBG myocardial scintigraphy in pre-motor Parkinson's disease: A review. *Parkinsonism Relat. Disord.*, 2013 Nov. 21. pii: S1353-8020(13)00393-3. doi: 10.1016/j.parkreldis.2013.11.001. [Epub ahead of print]PMID:24332912 [PubMed - as supplied by publisher].

[232] Yamamoto T., Uchiyama T., Sakakibara R., Taniguchi J., Kuwabara S. The subthalamic activity and striatal monoamine are modulated by subthalamic stimulation. *Neuroscience,* 2013 Nov. 27; 259C:43-52. [doi: 10.1016/j.neuroscience.2013.11.034. Epub ahead of print]PMID: 24291727[PubMed - as supplied by publisher].

[233] Sakakibara R., Tateno F., Yano M., Takahashi O., Sugiyama M., Ogata T., Haruta H., Kishi M., Tsuyusaki Y., Yamamoto T., Uchiyama T., Yamanishi T., Yamaguchi C. Imidafenacin on bladder and cognitive function in neurologic OAB patients. *Clin. Auton. Res.*, 2013 Jul. 3. [Epub ahead of print] PMID:23820664.

[234] Sakakibara R. Editorial Comment to Brain activity during bladder filling and pelvic floor muscle contractions: A study using functional magnetic resonance imaging and synchronous urodynamics. *Int. J. Urol.*, 2013 Jul. 1. doi: 10.1111/iju.12222. [Epub ahead of print] No abstract available. PMID:23815561.

[235] Yamamoto T., Sakakibara R., Uchiyama T., Yamaguchi C., Ohno S., Nomura F., Yanagisawa M., Hattori T., Kuwabara S. Time-dependent changes and gender differences in urinary dysfunction in patients with multiple system atrophy. *Neurourol. Urodyn.*, 2013 Jun. 11. doi: 10.1002/nau.22441. [Epub ahead of print] PMID:237544.

[236] Kazui H., Mori E., Ohkawa S., Okada T., Kondo T., Sakakibara R., Ueki O., Nishio Y., Ishii K., Kawaguchi T., Ishikawa M., Takeda M. Predictors of the disappearance of triad symptoms in patients with idiopathic normal pressure hydrocephalus after shunt surgery. *J. Neurol. Sci.*, 2013 May 15; 328(1-2):64-9.

[237] Aoki Y., Nakajima A., Sakakibara R., Ohtori S., Takahashi K., Nakagawa K. Pathologic thoracic spine fracture in presence of

Parkinson's disease and diffuse ankylosis: successful management of a challenging condition. *BMC Musculoskelet Disord.*, 2013 Feb. 11; 14:61. doi: 10.1186/1471-2474-14-61. PMID:23394219.

[238] Haruta H., Sakakibara R., Ogata T., Panicker J., Fowler C. J., Tateno F., Kishi M., Tsuyusaki Y., Uchiyama T., Yamamoto T. Inhibitory control task is decreased in vascular incontinence patients. *Clin. Auton. Res.*, 2013 Apr.; 23(2):85-9.

[239] Nakatsuka T., Imabayashi E., Matsuda H., Sakakibara R., Inaoka T., Terada H. Discrimination of dementia with Lewy bodies from Alzheimer's disease using voxel-based morphometry of white matter by statistical parametric mapping 8 plus diffeomorphic anatomic registration through exponentiated Lie algebra. *Neuroradiology,* 2013 May; 55(5):559-66.

[240] Sakakibara R. Editorial comment to Effective management of lower urinary tract dysfunction in idiopathic Parkinson's disease. *Int. J. Urol.*, 2013 Jan.; 20(1):84-5. doi: 10.1111/iju.12053. No abstract available. PMID:23281787.

[241] Sakakibara R., Kishi M., Tsuyusaki Y., Tateno A., Tateno F., Uchiyama T., Yamamoto T., Yamanishi T., Yano M. "Meningitis-retention syndrome": a review. *Neurourol. Urodyn.*, 2013 Jan.; 32(1):19-23.

[242] Tsunoyama K., Sakakibara R., Takahashi O., Sugiyama M., Uchiyama T., Tateno F., Kishi M., Tsuyusaki Y., Yamamoto T., Tanabe K. How the bladder senses? a five-grade measure. *LUTS,* 2013; 5: 17–22.

[243] Sakakibara R., Ito T., Yamamoto T., Uchiyama T., Yamanishi T., Kishi M., Tsuyusaki Y., Tateno F., Katsuragawa S., Kuroki N. Depression, anxiety and the bladder. LUTS: LOWER URINARY TRACT SYMPTOMS Article first published online : 31 MAR 2013, DOI: 10.1111/luts.12018.

[244] Sakakibara R. Editorial. Overactive bladder as a brain disease. *J. Neurological Disorders and Stroke*, 2013; 1: 1002 e-pub16th July.

[245] Sato M., Tsuruoka Y., Kuzuu A., Doi H., Masaka T., Sakakibara R. Monitoring blood l-dopa and l-dopa metabolite concentrations and adverse events in patients with advanced Parkinson's disease receiving l-dopa and MAO inhibitor selegiline ((-)deprenil) combination therapy: a clinical study. *Pharmacometrics (Oyo Yakuri),* 2013; 84: 41-46.

[246] Sato M., Kogure A., HAgino J., Doi H., Masaka T., Sakakibara R. Usefulness of monitoring blood l-dopa and l-dopa metabolite concentrations in patients with advanced Parkinson's disease receiving

COMT inhibitor entacapone combination therapy. *Pharmacometrics (Oyo Yakuri)*, 2013; 85: 15-24.

[247] Suzuki J., Sakakibara R., Tomaru T., Tateno F., Kishi M., Ogawa E., Kurosu T., Shirai K. Stroke and cardio-ankle vascular stiffness index. *J. Stroke Cardiovasc. Dis.*, 2013; 22: 171-175.

[248] Sakakibara R., Suzuki J., Tsuyusaki Y., Tateno F., Kishi M., Tomaru T. Stroke and cardio-ankle vascular stiffness index: a clinical use. A Review Article. *J. Neurological Disord. Stroke*, 2013: v1: 1111. E-pub.

[249] Tsuyusaki Y., Sakakibara R., Kishi M., Tateno F., Yoshida T. Downbeat nystagmus as the initial manifestation of anti-NMDAR encephalitis. *Neurol. Sci.*, 2014; 35: 125-126.

[250] Sakakibara R., Panicker J., Fowler C. J., Tateno F., Kishi M., Tsuyusaki Y., Yamanishi T., Uchiyama T., Yamamoto T., Yano M. Is overactive bladder a brain disease? The pathophysiological role of cerebral white matter in the elderly. *Int. J. Urol.*, 2014 Jan.; 21(1):33-8.

[251] Tsuyusaki Y., Sakakibara R., Kishi M., Tateno F., Aiba Y., Ogata T., Nagao T., Terada H., Inaoka T. "Invisible" Brain Stem Infarction at the First Day. *J. Stroke Cerebrovasc. Dis.*, 2014 May 6. pii: S1052-3057(14)00091-3.   doi: 10.1016/j.jstrokecerebrovasdis.2014.02.010. [Epub ahead of print] PMID:24809672 [PubMed - as supplied by publisher].

[252] Miyamoto S., Yoshimoto T., Hashimoto N., Okada Y., Tsuji I., Tominaga T., Nakagawara J., Takahashi J. C.; JAM Trial Investigators (Sakakibara R etc.). Effects of extracranial-intracranial bypass for patients with hemorrhagic moyamoya disease: results of the Japan Adult Moyamoya Trial. *Stroke,* 2014 May; 45(5):1415-21.

[253] Sakakibara R. Editorial Comment to Urinary incontinence in patients with Alzheimer's disease: Relationship between symptom status and urodynamic diagnoses. *Int. J. Urol.*, 2014 Mar 17. doi: 10.1111/ iju.12436. [Epub ahead of print] No abstract available.

[254] Suzuki J., Sakakibara R., Tateno F., Tsuyusaki Y., Kishi M., Ogata T., Tomaru T., Shirai K., Kurosu T. Parkinson's disease and the cardio-ankle vascular stiffness index. *Intern. Med., 2014*; 53(5):421-6.

[255] Sakakibara R., Tateno F., Nagao T., Yamamoto T., Uchiyama T., Yamanishi T., Yano M., Kishi M., Tsuyusaki Y., Aiba Y. Bladder function of patients with Parkinson's disease. *Int. J. Urol.*, 2014 Feb. 27. doi: 10.1111/iju.12421. [Epub ahead of print].

[256] Doi H., Sakakibara R., Sato M., Hirai S., Masaka T., Kishi M., Tsuyusaki Y., Tateno A., Tateno F., Takahashi O., Ogata T. Dietary herb

extract rikkunshi-to ameliorates gastroparesis in Parkinson's disease: a pilot study. *Eur. Neurol.*, 2014; 71(3-4):193-5.

[257] Doi H., Sakakibara R., Sato M., Hirai S., Masaka T., Kishi M., Tsuyusaki Y., Tateno A., Tateno F., Takahashi O., Ogata T. Nizatidine ameliorates gastroparesis in Parkinson's disease: a pilot study. *Mov. Disord.*, 2014 Apr.; 29(4):562-6.

[258] Sakakibara R., Tateno F., Kishi M., Tsuyusaki Y., Terada H., Inaoka T. MIBG myocardial scintigraphy in pre-motor Parkinson's disease: a review. *Parkinsonism Relat. Disord.*, 2014 Mar.; 20(3):267-73.

[259] Yamamoto T., Uchiyama T., Sakakibara R., Taniguchi J., Kuwabara S. The subthalamic activity and striatal monoamine are modulated by subthalamic stimulation. *Neuroscience,* 2014 Feb. 14; 259:43-52.

[260] Sakakibara R., Tateno F., Yano M., Takahashi O., Sugiyama M., Ogata T., Haruta H., Kishi M., Tsuyusaki Y., Yamamoto T., Uchiyama T., Yamanishi T., Yamaguchi C. Imidafenacin on bladder and cognitive function in neurologic OAB patients. *Clin. Auton. Res.,* 2013 Aug.; 23(4):189-95.

[261] Sakakibara R. Editorial comment to Brain activity during bladder filling and pelvic floor muscle contractions: a study using functional magnetic resonance imaging and synchronous urodynamics. *Int. J. Urol.,* 2014 Feb.; 21(2):174.

[262] Yamamoto T., Sakakibara R., Uchiyama T., Yamaguchi C., Ohno S., Nomura F., Yanagisawa M., Hattori T., Kuwabara S. Time-dependent changes and gender differences in urinary dysfunction in patients with multiple system atrophy. *Neurourol. Urodyn.,* 2014 Jun.; 33(5):516-23.

B. Book Chapters

[1] Sakakibara R., Fowler C. J.: Cerebral control of bladder, bowel, and sexual function and effects of brain disease In Fowler C. J. editor. Neurology of bladder, bowel, and sexual function Boston: Butterworth-Heinemann, 229-243 1999.

[2] Sakakibara R.: Parkinsonian disorders and pure autonomic failure (Chapter 8) In Fowler C. J. editor. Seminars in Clinical Neurology (by World Federation of Neurology) Neurologic bladder, bowel, and sexual function Boston: Elsevier 85-92 2001.

[3] Sakakibara R., Fowler C. J.: Brain disease (Chapter 9) In Fowler C. J. editor. Seminars in Clinical Neurology (by World Federation of

Neurology) Neurologic bladder, bowel, and sexual function Boston: Elsevier 93-109 2001.

[4]   Sakakibara R., Hattori T.: Dementia and lower urinary tract dysfunction (Chapter 21) In Corcos J, Schick E editors. Neurogenic Bladder: Adults and Children London: Martin Dunitz Publisher 245-257 2004.

[5]   Sakakibara R., Fowler C. J., Hattori T.: Urinary dysfunction in multiple system atrophy (Chapter 23) In Corcos J., Schick E. editors. Neurogenic Bladder: Adults and Children London: Martin Dunitz Publisher 265-273 2004.

[6]   Sakakibara R., Hattori T.: Dementia and lower urinary tract dysfunction (Chapter 20) In Corcos J, Schick E editors. Neurogenic Bladder: Adults and Children, 2nd edition, London: Informa Healthcare, 256-273 2008.

[7]   Sakakibara R., Fowler C. J., Hattori T.: Urinary dysfunction in multiple system atrophy (Chapter 22) In Corcos J., Schick E. editors. Neurogenic Bladder: Adults and Children, 2nd edition, London: Informa Healthcare, 282-293 2008.

[8]   Wyndaelle J. J., Kovindha A., Madesbacher H., Radziszewski P., ruffion A., Schurch B., Castro D., Igawa Y., Sakakibara R., Perkash I.: Neurologic urinary and fecal incontinence, In: Incontinence, 4th edition, Editors: Abrams P., Cardoso L., Khoury S., Wein A., International Consultation on Incintinence, pp797-960, 2009.

[9]   Sakakibara R., Fowler C. J., Hattori T. Parkinson's disease. In: Pelvic Organ Dysfunction in Neurological Disease: Clinical Management and Rehabilitation. Eidtors: Chaudhuri Fowler C. J., Panicker J., Emmanuel A., Cambridge University Press, UK, 2010.

[10]   Sakakibara R., Fowler C. J., Hattori T. Multiple system atrophy. In: Pelvic Organ Dysfunction in Neurological Disease: Clinical Management and Rehabilitation. Eidtors: Fowler C. J., Panicker J., Emmanuel A., Cambridge University Press, UK, 2010.

[11]   Sakakibara R, Fowler C. J., Hattori T. Cerebral diseases. In: Pelvic Organ Dysfunction in Neurological Disease: Clinical Management and Rehabilitation. Eidtors: Fowler C. J., Panicker J., Emmanuel A., Cambridge University Press, UK, 2010.

[12]   Sakakibara R. Basal ganglia circuit and bladder function: clinical implication to Parkinson's disease. In: Integrated Circuits, Photodiodes and Organic Field Effect Transistors. Editors: Robert McIntire and Pierre Donnell. NOVA Publishers 2010; pp. 369-382.

[13]   Sakakibara R., Uchiyama T., Yamamoto T., Kishi M., Ogawa E., Tateno F. Sexual problems in Parkinson's disease. In: Psychiatry of Parkinson's

disease. Editors: Ebmeier K. P., O'Brien J. T., Taylor J. P. Adv. Bol. Psychiatry, series No. 27, Karger, Basel, 2012: 71-76.

[14] Sakakibara R., Uchiyama T., Yamamoto T., Tateno F., Yamanishi T., Kishi M., Tsuyusaki Y. Chapter 12: Sphincter EMG for diagnosing multiple system atrophy and related disorders. In: Computational intelligence in electromyography analysis: a perspective on current applications and future challenges. Editor: Naik G. R. *Adv. Bol. Psychiatry,* series No. 27, Intech Publisher, New York, 2012: 287-306.

[15] Sakakibara R., Kishi M., Tsuyusaki Y., Tateno F., Uchiyama T., Yamamoto T., Suzuki Y. Chapter 2: Constipation in Parkinson's disease. In: Constipation and irritable bowel syndrome. Editors: Kiyomizu G., Rin H. Nova Science Publishers Inc., 2012: 51-83.

[16] Sakakibara R., Tsuyusaki Y., Tateno F., Kishi M., Uchiyama T., Yamamoto T. Chapter 3: Constipation with low myocardial metaiodobenzylguanidine uptake are the initial features of Parkinson's disease. In: Hydrocephalus: symptoms, treatment and potential complications. Editors: Yoshida C., Ito A. Nova Science Publishers Inc., 2012: 217-222.

[17] Sakakibara R., Tateno F., Kishi M., Tsuyusaki Y., Uchiyama T., Yamanishi T., Yamamoto T. Chapter 3: Normal-pressure hydrocephalus: a common sourse of elderly incontinence with brain etiology. In: Hydrocephalus: symptoms, treatment and potential complications. Editors: Velazquez A. Nova Science Publishers Inc., 2013: 85-96.

In: Horizons in Neuroscience Research. Vol. 25   ISBN: 978-1-63485-286-9
Editors: A. Costa and E. Villalba   © 2016 Nova Science Publishers, Inc.

Chapter 8

PURE MOTOR MONOPARESIS DUE TO ISCHEMIC STROKE

Akiyuki Hiraga, MD*
Department of Neurology, Chiba Rosai Hospital,
Chiba, Japan

ABSTRACT

Pure motor monoparesis (PMM), an isolated paresis due to ischemic stroke in a single arm or leg without higher, cranial, or sensory dysfunction, is an uncommon clinical condition. Cortical infarctions of the precentral knob and anterior cerebral artery territory are the most commonly reported PMM lesions affecting the upper and lower extremities, respectively. Other reported lesion sites include the subcortex, corona radiata, internal capsule, and brainstem; however, they are reported less frequently than cortical infarctions because the descending motor axons in these areas are highly compacted. Large artery atherosclerosis and cardiogenic embolism are considered as the two main etiologies of cortical infarction. In earlier studies, PMM cases due to lacunar infarctions were presumed to be rare. However, lacunar infarction PMM cases have been increasingly reported in the recent years. The weakness patterns in PMM are complex and include proportional and distal- or proximal-dominant weakness. Distal-dominant weakness is the

---

* Address correspondence to: Akiyuki Hiraga, MD, Department of Neurology, Chiba Rosai Hospital, 2-16 Tatsumidai-Higashi, Ichihara-shi, Chiba 290-0003, Japan. Telephone number: +81-436-74-1111, Fax number: +81-436-74-1151, e-mail: hiragaa@yahoo.co.jp.

most common PMM type. Some patients exhibit isolated finger paresis or predominant weakness of specific fingers. Although the overall prognosis is favorable in PMM due to ischemic stroke, short- and long-term prognosis depends on risk factors and stroke mechanisms. The diagnosis is sometimes difficult because many cases of PMM lack pyramidal signs and mimic peripheral nerve damage. Therefore, acute monoparesis, particularly in elderly patients with conventional risk factors, should be carefully assessed. Diffusion-weighted imaging is the most useful diagnostic tool for PMM due to ischemic stroke as small cortical infarctions often cannot be detected by conventional sequences of magnetic resonance imaging, or computed tomography. This review provides an update on clinical features, lesion topography, and prognosis of PMM due to ischemic stroke.

Keywords: pure motor monoparesis, cerebral infarction, diffusion weighted magnetic resonance imaging

INTRODUCTION

Pure motor monoparesis (PMM) of a single arm or leg due to ischemic stroke (IS) without higher, cranial, or sensory symptoms/disturbances is a rare condition [1, 2]; however, it is a clinically important stroke syndrome that can easily be misdiagnosed and/or confused with nonvascular causes of weakness [3, 4]. Undoubtedly, correct diagnosis of acute stroke is of paramount importance to clinicians to enable selection of appropriate treatments and to ensure prevention of acute complications, including recurrent stroke [5]. Given the rarity of PMM due to IS, which often mimics peripheral neuropathy and is sometimes difficult to detect by computed tomography (CT) or conventional magnetic resonance imaging (MRI), many patients with PMM due to IS might have been overlooked until recently. The diagnosis of PMM due to IS has always been a clinical challenge not only for general practitioners but also for neurologists. This review provides an update on the frequency, lesion topography, clinical features, mechanisms, and prognosis of PMM due to IS. Moreover, this review aims to aid clinicians in improving the accurate diagnosis of PMM due to IS, with a focus on differential diagnosis from nonvascular causes of PMM, especially from peripheral nerve damage.

A Brief History of PMM Due to IS

Lhermitte described "pseudoperipheral palsy" in 1909 in patients with a sensorimotor deficit in fingers due to central nervous system lesions [6]. In 1986, Ashizawa et al. reported one patient with ipsilateral leg monoparesis and four patients with ipsilateral arm monoparesis due to brain tumors or brain abscesses located in the contralateral superficial cerebral hemisphere [7], suggesting that monoparesis of the arm or leg due to brain damage was caused by a space-occupying lesion. Between the 1980s and early 1990s, several reports showed superficial infarctions presenting as PMM [8-13]. In 1997, based on a functional MRI study, cortical representation of the hand was determined to be located in the knob-like structure within the precentral gyrus (motor cortex), which was termed the precentral knob; that report also presented a case with hand monoparesis and sensory disturbance due to a precentral knob infarction [14]. Subsequently, numerous studies of hand PMM or monoparesis with sensory disturbances due to precentral knob infarctions were reported. Moreover, diffusion-weighted imaging (DWI), which provides accurate lesion location, led to more frequent reporting of PMM due to IS. In 2005, two large studies from Italy and Switzerland reported the frequency and lesion topography of PMM due to stroke [1, 2]. In 2011, the first English review of PMM due to IS was published from Japan [4].

PMM Criteria

Two large studies [1, 2] of PMM due to acute stroke outlined the criteria for PMM as follows: (1) first acute stroke (ischemic or hemorrhagic), (2) a motor deficit in a single body part (face, arm, or leg), (3) no sensory deficit (by subjective complaints and clinical examination), and (4) no visual field deficit. The criteria established by the second study are: (1) limb paresis without any sensory disturbance, coordination or language deficit, or significant involvement of speech or ipsilateral face, and (2) motor deficit caused by ischemic or hemorrhagic stroke, confirmed both clinically and radiologically. The criteria for distal arm PMM due to IS are isolated weakness of one arm or hand that is not associated with objective sensory, coordination, or language deficits, and no significant dysarthria or ipsilateral face or leg weakness [15]. Based on these criteria, PMM due to IS is defined as the presence of an isolated motor deficit in a single leg or arm with no sensory symptoms/disturbances or coordination deficits and no significant involvement of speech or ipsilateral face; the presence of a motor deficit caused by IS is confirmed both clinically and radiologically [4]. Brachiofacial motor stroke [16, 17] is excluded due to facial involvement. A number of previously reported cases of monoparesis due to IS had sensory disturbances, thus they do not fulfill the abovementioned current criteria for PMM due to IS. However, some reports of monoparesis due to IS, especially those with cortical infarctions, include both PMM and monoparesis with sensory symptoms and are often difficult to clearly distinguish. Some of these cases are described below.

FREQUENCY

The reported frequency of PMM due to IS is summarized in Table 1. Earlier studies using CT imaging did not always detect causative PMM lesions; thus, the exact frequency of PMM due to IS is difficult to assess [4]. Previous reports including PMM due to hemorrhagic stroke showed that PMM of the leg was rarer than that of the arm (14% vs. 86% [11], 28% vs. 72% [1], 33% vs. 67% [2]). However, one DWI study showed that the frequencies of leg and arm PMM were 54% and 46%, respectively, suggesting that leg PMM might not always be less common than arm PMM [4]. Most studies did not examine the differences in frequencies of arm and leg PMM as they primarily focused only on PMM cases involving the arm, focusing on precentral knob infarctions, and did not include leg PMM patients. Hemorrhagic stroke can also cause PMM [1, 2, 11, 13, 18-20], albeit at a lower rate than IS. As an example, Figure 1A shows traumatic intracranial hemorrhage presenting as leg PMM.

LESION TOPOGRAPHY AND CLINICAL FEATURES

IS of the cortical surface is more likely to cause PMM than deeper brain lesions, which frequently cause pure motor hemiparesis; this is due to the compaction of descending motor axons in a small area in the subcortex, internal capsule, and brainstem [4]. In this regard, cortical infarction of the hand motor area is the most frequent site affected by arm PMM, whereas cortical infarction of the anterior cerebral artery (ACA) territory is the most frequently reported site involved in leg PMM [1, 2].

Table 1. Reported frequency of pure motor monoparesis due to ischemic stroke

References	Year	Modality	Frequency and lesion locations
Nelson et al. [107]	1980	CT	1 (1 arm) of 26 pure motor stroke. Lesion: no visual lesion.
Donnan et al. [86]	1982	CT	1 (1 leg) of 69 pure motor stroke. Lesion: posterior limb of internal capsule.
Rascol et al. [85]	1982	CT	1 (1 leg) of 30 pure motor stroke. Lesion: internal capsule infarction widespread to putamen and caudate nucleus.
Weisberg [8]	1984	CT	4 (leg or arm is not described) of 40 patient with small superficial cerebrovascular lesion.
Arboix et al. [87]	1990	CT/MRI	8 (4 arm and 4 leg) of 227 patients with lacunar infarctions (8 of 125 pure motor stroke). Lesion: 1 of 8 had internal capsule (arm or leg is not described) 7 had no visual lesion.
Boiten and Lodder [10]	1991	CT	7 (5 arm and 2 leg) of 252 patients with first-ever supratentorial ischemic stroke (2 had slight sensory disturbance). Lesion: 6 of 7 had cortical infarctions. 1 patient had no visual lesion.
Chimowitz et al. [105]	1991	CT/MRI	11 (leg or arm is not described) pure motor monoparesis of 81 patients with pure motor, sensorimotor, or pure sensory stroke (11 of 48 pure motor stroke). Lesion: this report showed no details of lesion topography.
Chamorro et al. [78]	1991	CT/MRI	2 of (2 arm) 337 patients with lacunar stroke (sensory disturbance of these 2 patients were not described).
Melo et al. [11] *	1992	CT/MRI	29 (25 arm and 4 leg) of 255 acute stroke patients.

Table 1. (Continued)

			Arm lesion: deep infarcts 2, superficial infarcts 12, hemorrhagic stroke 1, and no visible lesion 10.
			Leg lesion: deep infarcts 1, superficial infarcts of ACA 2, no visible lesion 1.
Mohr et al. [12]	1993	CT	8 (all arm) cases of 183 with cerebral convexity infarction of the MCA territory. Lesion: mid-Rolandic region.
Castaldo et al. [15]**	2003	CT/MRI	35 arm of 4818 ischemic stroke patients. Lesion: all had MCA territory infarction including parietal lobe.
Maeder-Ingvar et al. [1]*	2005	CT/MRI	153 (123 arm, 30 leg) of 4802 stroke patients.
			Arm lesion: MCA 59%, subcortex 26%, brainstem 9%, ACA 2%, others 4%.
			Leg lesions: ACA 43%, subcortex 27%, MCA 23%, brainstem 7%.
Paciaroni et al. [2]*	2005	CT/MRI	51 (34 arm, 17 leg) of 2003 stroke patients.
			Arm lesion: subcortical (corona radiata, centrum semiovale) 35.3%, cortical branches of anterior superficial MCA 26.5%, deep perforators of MCA 23.5%, thalamocapsular 8.8%, and cortical branches of posterior superficial MCA 5.9%.
			Leg lesion: cortical branches of the ACA 23.5%, watershed infarcts 17.6%, thalamocapsular 17.6%, striatocapsular 11.8%, pons 11.8%, subcortical (corona radiata, central semiovale) 5.9%, deep perforators of MCA 5.9%, and cortical branches of the anterior superficial MCA 5.9%.
Celebisoy et al. [44]**	2007	MRI	7 hand of 815 patients who were hospitalized in the stroke care unit (this report did not described sensory symptoms).
			Lesion: posterior portion of the precentral gyrus.
Alstadhaug and Sjulstad [40]***	2013	MRI/CT	13 hand of 866 ischemic stroke patients (15 had hand paresis and remaining 2 of 15 had sensory disturbance, not PMM). Lesion: this study was limited to hand motor area.

CT, computed tomography: MRI, magnetic resonance imaging; ACA, anterior cerebral artery; MCA, middle cerebral artery.
*These studies included hemorrhagic stroke.
**These studies focused only on upper limb monoparesis, and the research criteria did not include leg monoparesis.
***This study focused only on hand paresis due to hand knob area infarction, and other topographies or leg monoparesis were not included. In this study, only one patient underwent CT, and the remaining patients underwent diffusion weighted imaging.
Note: 34 of 35 patients underwent MRI in the study by Castaldo et al. [15].

Table adapted from Hiraga [4].

Additionally, isolated distal, proximal, and distal-predominant monoparesis cases are observed more frequently in cortical PMM than in subcortical PMM. Selective damage to primary motor cortex results in restricted paresis of the contralateral upper and lower limbs in patients with cortical PMM. Pyramidal signs vary in each case but are relatively infrequently seen in both cortical and subcortical cases of PMM due to IS [4].

This table does not contain the study by Timisit et al. [21], which reported five patients with hand palsy among 690 patients hospitalized in the stroke care unit. All five had sensory disturbances. This table also does not include the study by Peters et al. [35] that was limited to hand knob area infarctions presenting with hand paresis and used diffusion-weighted imaging for all patients; they found that 29 of 3499 patients treated in the stroke care unit had hand paresis and that 22 of 29 patients showed isolated hand paresis. However, 10 of these 29 patients showed sensory disturbances; thus, the exact frequency of PMM was not determined.

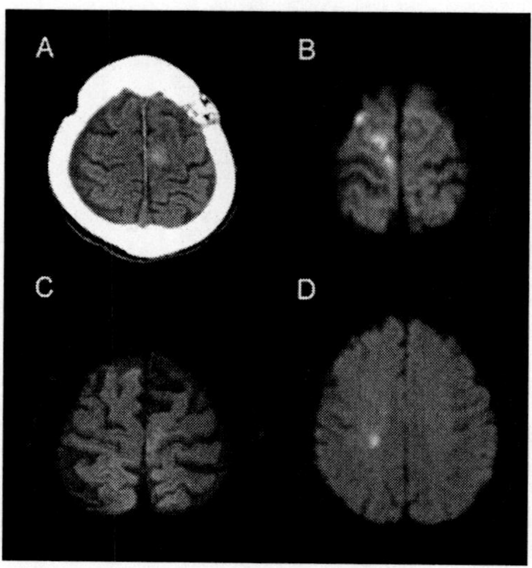

Figure 1. Neuroimaging studies of patients with pure motor monoparesis (PMM) of lower extremities. (A) Computed tomography of a patient with PMM of the right leg due to traumatic intracerebral hemorrhage located in the left leg motor area. (B) Diffusion-weighted images (DWI) of a patient with PMM of the left leg due to acute cerebral infarction located in the right anterior cerebral artery (ACA) territory. (C) DWI of a patient with PMM of the right leg showing acute infarction located in the left ACA territory. This case presented with monoparesis restricted to the foot. (D) DWI of a patient with left leg PMM due to acute cerebral infarction located in the right central semiovale (D).

Cerebral Cortex

Arm

Superficial [8, 10, 13] and cerebral convexity infarctions (mid-Rolandic lesions) [12], infarctions of the anterior and posterior superficial middle cerebral artery (MCA) territories [2, 11], and infarctions of the pial branches of MCA [9] have been reported as causes of upper limb PMM.

The areas affected in PMM or upper limb monoparesis due to cortical infarction are the precentral gyrus, especially the precentral knob, and the parietal lobe; the former is more frequently reported. One study reported six patients with isolated hand palsy due to IS of white matter of the angular gyrus [21]. All patients had sensory symptoms/disturbances, and one had aphasia; none fulfilled the criteria for PMM, suggesting that the parietal lobe was involved in motor control of the hand. Another study of 35 upper limb PMM cases reported that all patients had infarctions in the MCA territory [15]; nine of the 35 cases had some involvement of both the precentral and postcentral gyri, and 17 had some postcentral gyrus involvement. Importantly, all patients exhibited some involvement of the parietal lobe infarct, indicating that the parietal lobe was important for PMM development. The inferior parietal lobe may also be important in PMM [22].

Numerous studies showed precentral knob infarctions presenting as PMM or upper limb monoparesis [22-42]. Some reports found that the causative lesions were in the posterior portion of the precentral gyrus (i.e., the anterior bank of the central sulcus) [30, 32, 43, 44], whereas others indicated the frontal lobe, precentral gyrus [45, 46], or motor cortex [47]. Most studies showed that patients with upper limb PMM due to precentral gyrus infarctions had either distal or distal-dominant weakness, whereas several cases presented with proportional weakness [13]. In a study of 12 patients with predominant weakness in specific fingers and small cortical infarct, two of whom had PMM [29], MRI revealed that the lesions were related to predominant involvement of ulnar fingers and were located significantly more medially than those associated with predominant involvement of radial fingers in the presumed hand representation area of the motor cortex. These observations support traditional views [48], although individual variations in motor topography were also found [29, 31]. Others showed that ulnar or radial fingers were predominantly affected in monoparesis due to cortical infarctions [22, 23, 28, 30, 32, 43, 44, 46]. Furthermore, some studies indicated ulnar–medial and radial–lateral correlations [28, 32, 43]; patients with large lesions had uniform finger involvement [28]. Likewise, PMM due to cortical infarctions, mimicking radial or ulnar nerve palsy, a condition resulting from precentral knob infarction that imitates anterior interosseous nerve palsy, was also reported [37].

In 1999, Schieber and colleagues emphasized that they identified no cases with individual digit paresis due to IS in which the greatest weakness was in the index, middle, or ring fingers [43], providing little evidence that individual digits were represented in separate cortical territories; instead, these findings suggested broadly overlapping (i.e., lateromedial) gradients in the cortical representation of radial and ulnar fingers. Although cases with isolated thumb [34] or index finger paralysis [31, 41] and those with isolated ring finger palsy [38] due to precentral knob infarctions were reported, there are no reported cases with isolated middle finger or little finger PMM due to IS. However, predominant thumb weakness was noted in some cases; one patient with significant thumb weakness during the recovery phase of hemiparesis exhibited predominant thumb flexion weakness [49], whereas another patient with isolated fingers weakness due to precentral gyrus infarction had predominant thumb weakness wherein thumb adduction and flexion were strongly affected [23]. These two cases, in addition to an isolated complete thumb paralysis case [34], provide supporting evidence that motor hand areas might contain a subregion that selectively represents the thumb [43]. Moreover, two previously reported cases with isolated index finger weakness [31, 41] and one patient with prominent index finger weakness sparing the thumb and largely sparing the ulnar fingers [33] suggest the presence of a subregion selectively representing the index finger.

Although most cases of upper limb PMM due to cortical infarctions present with distal-dominant weakness, cases of isolated shoulder paresis were also reported [4, 50-54]. The first reported isolated shoulder PMM case in 2003 had an infarcted region restricted to the base of the central sulcus and in the deep white matter, located midway between the precentral knob and longitudinal cerebral fissure [50]. One patient with an isolated shoulder seizure due to a small cortical infarction had a lesion in a similar region [55], indicating that this location correlated with human shoulder motor somatotopy. Cases with PMM limited only to shoulder and arm muscles also had precentral gyrus infarctions in the precentral gyrus, medially to the precentral knob [56]. Although patients presenting with an affected wrist flexor were also diagnosed with PMM [25], predominant wrist extensor involvement was more frequently observed [26, 27, 44]. Furthermore, not many differences between finger extensors and flexors occur with precentral knob infarctions, except for mild differences observed in some cases [30, 44]. Deep tendon reflexes (DTRs) of the affected limb in PMM cases with precentral knob infarction were either normal or symmetric in some patients [22, 27, 37]. In particular, normal or symmetric DTRs were evident in patients with weakness in isolated or specific fingers [25, 31, 33, 34], as well as in those with isolated shoulder weakness [50, 51, 53]. However, several patients exhibiting weakness in specific fingers [29] and one patient with isolated shoulder palsy [54] had active DTRs of the affected limb. Other case reports also showed active or brisk DTRs [24, 28, 42, 43]. The Babinski sign was negative in most cases; thus far, only one case presenting with a positive Babinski sign has been reported [22]. In one study on cortical infarction presenting as arm PMM, patients demonstrated all DTR patterns: symmetrical, active, or decreased [13]. A case of claw hand deformity in acute stage of PMM due to hand knob area infarction mimicking peripheral nerve disease has been reported [42], although the majority of stroke-associated PMM cases did not exhibit any hand deformities in the acute stage. The differences in PMM and monoparesis with sensory disturbances might partially be due to postcentral gyrus involvement, as those with infarct regions involving both the precentral and postcentral gyri had sensory symptoms [25, 32]. Figures 2A and 2B show a patient with precentral knob infarction presenting as isolated hand PMM. Figure 2C shows a patient with radial-dominant hand PMM due to a precentral knob infarction, and Figure 2D shows a patient with proximal-dominant upper limb PMM due to cortical infarction.

Leg

Cortical branches of the ACA are the most common sites affected in leg PMM [2]. Classically, leg-predominant weakness due to stroke results from ACA territory infarctions and consequent paracentral lobule involvement [57]. Leg PMM due to cortical infarctions occurs in the precentral gyrus [58-67], paracentral lobe [68-73], and supplementary motor area [74]. Other small cortical infarctions in the ACA territory were also reported [10, 11]. Although the details of infarcts were not described, four patients with leg PMM due to occlusion or severe stenosis of internal carotid artery were reported [75].

The weakness patterns of leg PMM due to ACA territory infarctions can be proportional [68, 69], distal-dominant [59-62, 64-67, 72, 74], or proximal-dominant [59, 63, 73]. The rear portion of the medial precentral gyrus is the site affected in distal-dominant weakness [59]. One patient was reported to develop isolated toe paralysis in the recovery phase of a paracentral lobule infarction [72]. Additionally, several cases presenting with isolated foot drop due to precentral gyrus infarctions were reported [62, 64-67, 74]. These reports

indicate that a small ACA territory infarction may cause isolated distal leg
PMM. Some patients were reported to exhibit active DTRs of the affected leg
[60,64,69], whereas normal or symmetric DTRs were observed in other cases
[61-68, 72-74]; one patient with a decreased DTR was also reported [61]. A
positive Babinski sign on the affected side was demonstrated in several cases
[64, 69, 72, 73], whereas it was negative in others [60-65, 67, 74]. Figures 1B
and 1C show ACA territory cortical infarctions presenting as leg PMM; the
case illustrated in Figure 1C presented with isolated foot paresis.

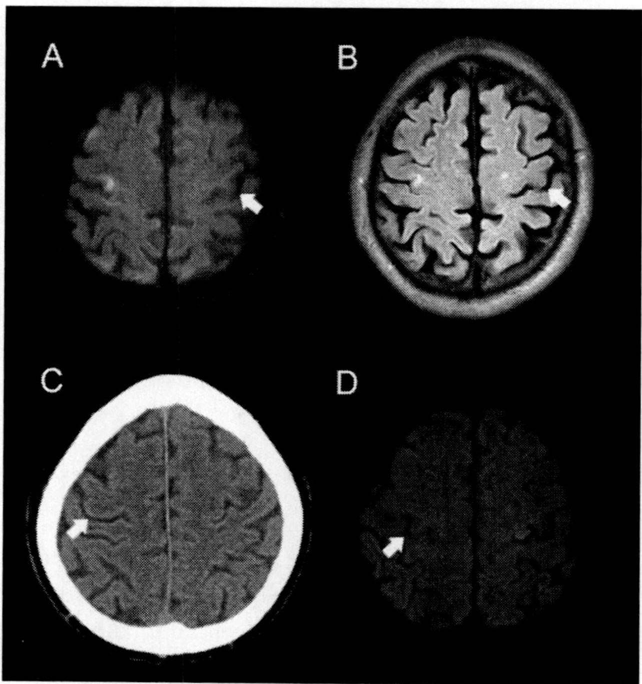

Figure 2. Neuroimaging studies of patients with pure motor monoparesis (PMM) of
upper extremities. (A) Diffusion-weighted images (DWI) of a patient with PMM of the
left hand due to an acute cerebral infarction located in the hand motor area. (B) Fluid-
attenuated inversion recovery images showing an infarction of the same area. (C)
Computed tomography of a patient with PMM of the right hand. This case presented
with predominant involvement of radial fingers, and the affected region was located in
the lateral region of the presumed hand representation area of the left motor cortex
(precentral knob). This case indicated ulnar-medial and radial-lateral correlations. (D)
DWI of a patient with proximal dominant-PMM of the right upper extremity, showing
an infarction located in the left precentral gyrus medial to the knob-like structure, and
laterally to the shoulder motor area. Arrows indicate the contralateral precentral knob.

Subcortex, Centrum Semiovale, and Corona Radiata

Although some patients with cortical infarctions presenting as PMM also had adjacent subcortical regions that were infarcted, a subcortical infarction without affected cortical regions can also lead to arm or leg PMM. A large study showed that 26 patients with arm PMM and 27 patients with leg PMM had subcortical infarctions [1]. Another large study showed that ten patients with subcortical (corona radiata or centrum semiovale) infarctions had PMM; most cases had PMM of the arm [2]. A report on subcortical infarctions of the superficial MCA territory affecting the centrum semiovale showed that 2 of the 36 patients had upper limb monoparesis [76]. Other reports presented cases with frontal subcortical infarctions in the ACA territory that presented as leg PMM [68, 77]. A case report showed that brachial monoparesis in a patient was due to a corona radiata infarction [78]. Another study including eight arm PMM and six leg PMM patients [79] noted that patients with leg PMM had more posteriorly and medially located infarct regions than those with bulbofacial paresis or arm PMM, indicating anterolateral to posteromedial distribution. Other studies also indicated anteroposterior somatotopy in the corona radiata; leg fibers were located more posteriorly than arm fibers [68, 80, 81]. One diffusion tensor tractography (DTT) study showed that the hand fiber tract was located in the anteromedial portion of corona radiata [82], while another DTT report demonstrated that hand somatotopy of the corticospinal tract (CST) was located anterolaterally to the leg somatotopy of the CST in central semiovale [83]. These DTT findings and PMM case studies indicate that hand fibers are located more anteriorly than leg fibers and that anterior corona radiata infarction might manifest as hand PMM. PMM involving one or more fingers due to subcortical infarction was rarely reported. However, a patient with palsy of only ring and small fingers due to a subcortical infarction was reported [84]. Moreover, another patient had profound weakness and decreased sensation in the thumb and index finger due to a corona radiata infarction [46]; this patient was not diagnosed with PMM due to accompanying sensory disturbances. However, these observations indicate that involvement of specific fingers can occur not only due to a cortical infarction, but also due to a subcortical or corona radiata infarction, suggesting finger somatotopy might also exist in the subcortical region.

Patients with ACA territory subcortical infarctions and distal-dominant [77] or proportional [68] weakness were reported to have normal DTRs. Additionally, cases with leg PMM due to corona radiata infarctions had proportional weakness with brisk or normal DTRs of the affected leg [68]. Figure 1D shows a subcortical infarction presenting as leg PMM.

Internal Capsule

A CT study showed that leg PMM could result from an internal capsule infarction spreading to the putamen and caudate nucleus [85] or from an infarction located just lateral to the posterior limb of the internal capsule [86]. In another report, PMM due to an internal capsule infarction was detected by MRI [87], although it was not clear whether the arm or the leg was affected. Yet another report suggested that isolated arm monoparesis might also be due to infarction of the internal capsule [78]; however, no information was enclosed regarding sensory disturbances in that study. Other cases involving infarction of the posterior limb of the internal capsule presenting as leg PMM were also reported [59, 68]. Moreover, one report described a patient with infarction of the posterior limb of the internal capsule who developed an abrupt fall but showed only unilateral pyramidal tract signs in lower extremities during neurological examination [88].

A previous large study on monoparesis found that internal capsule infarctions, either striatocapsular or thalamocapsular, led to leg PMM in a small number of patients [2]. Traditionally, the somatotopic CST organization in the internal capsule is a homunculus, i.e., the hand fibers are located more anteriorly and slightly medially to the foot fibers. However, considerable overlap was demonstrated between hand and foot fibers [89]. A linear infarction along the long axis of the internal capsule was characteristic of posterior limb internal capsule infarction presenting as leg PMM [68]. These two cases support the results of an MR tractography study [90, 91], which demonstrated that CST foot fibers in the posterior limb of internal capsule might be somatotopically posteromedial to the hand fibers along the short axis of internal capsule. However, other studies showed that hand fibers were located anteromedial to foot fibers in the posterior limb of internal capsule [82, 92].

Clinically, two reported cases with leg PMM due to internal capsule lesions had slight proximal-dominant weakness with no Babinski sign and normal DTRs [68], whereas a separate patient had proportional leg weakness [59].

Brainstem

An acute brainstem infarction is difficult to detect by CT; thus, few brainstem infarctions presenting as PMM by CT were reported. One earlier CT study presented a patient with arm PMM due to a pontine infarction [93], whereas a larger study showed only two leg PMM cases resulting from pontine infarctions, with no patients presenting exhibiting arm PMM [2]. Another study showed nine arm PMM and seven leg PMM cases due to brainstem infarctions [1]. Distal-dominant brachial monoparesis results from infarction of the ventromedial pontine base on DWI, indicating that arm fibers are located more medially than leg fibers [94]. Rostral lateral pontine infarction presenting with leg PMM in three patients [95] indicated that the lesions responsible for crural monoparesis were presumably located in the dorsal and lateral regions of pontine base. Another study indicated that the lesions in PMM patients with leg-predominant weakness were dorsal and lateral in the pons [59]. The regions for pontine infarctions that did not present with PMM were located in the ventromedial region in patients with lower predominant hemiparesis, and were correlated with lesions in the dorsomedial and dorsolateral portions in patients with upper predominant hemiparesis [80], whereas the upper extremity fibers were located more ventromedially than the lower extremity fibers in four patients with brachial monoparesis and dysarthria due to pontine base infarctions [96]. A voxel-based MRI analysis of the basilar pons [97] indicated that upper extremity weakness was correlated with specific pontine regions; however, another study [98] did not show significant topographical differences between patients with paresis predominantly affecting the arm or leg, suggesting that the topographical arm/leg distribution was disrupted in the pons and that PMM due to brainstem infarctions was rare. However, DTT studies indicated the presence of motor somatotopy of the hand and leg in the brainstem. Specifically, hand somatotopy of the CST is located in the anteromedial portion of pons and leg somatotopy of the CST is located posterolateral to hand somatotopy of the CST [82, 99]. Additionally, hand somatotopy of the CST is located in the medial portion of the medullary pyramid, whereas leg somatotopy of the CST is situated in the lateral portion of the medullary pyramid [100]. The case report also suggests that the CST fibers innervating lower and upper extremities decussate at different levels in the lower medulla oblongata [101].

PMM due to midbrain or medulla infarctions is extremely rare. There are several case reports of patients with cerebral peduncular or medial medullary infarction presented with monoparesis; however, these cases either had accompanying sensory symptoms or did not have detailed description of symptoms [59, 102, 103]. One Japanese study reported a patient with PMM due to a medullary infarction that was limited to the upper lateral region of medullary pyramid [104]. PMM due to lateral medullary infarction was also reported by a group from Japan [101].

Assessment of clinical features reveals that patients with arm PMM due to pontine infarctions develop proportional weakness with normal DTRs [93] or distal-dominant arm weakness with active DTRs [94], both in the absence of a Babinski sign. Among patients with leg PMM due to rostral lateral pontine infarctions, two cases with dorsolateral infarcts were proximally dominant, and one case with an anterolateral infarct had proportional weakness [95]. One patient with a medullary infarction presenting as leg PMM showed proportional weakness with active DTRs [104], and two patients with leg PMM due to medullary infarctions had a positive Babinski sign in the affected leg [101, 104].

MECHANISMS OF STROKE IN PMM

PMM frequently occurs following cortical infarctions, and findings from earlier reports suggest that PMM rarely results from lacunar infarctions [11, 13, 105]. Eleven patients with leg or arm monoparesis revealed that the etiology of stroke was occlusive large artery disease and cardioembolism in six and three cases, respectively; the cause was either undetermined or miscellaneous in the remaining two patients [105]. Melo et al. reported 28 patients with PMM due to IS; the causes were determined as deep infarcts in 3 patients, superficial infarcts in 14 patients, and no visible lesions in 11 patients [11]. In contrast, a large study indicated that the etiologies of PMM due to IS were small artery disease, cardioembolism, and atherosclerosis in 39.2%, 15.7%, and 9.8% of patients, respectively [2], suggesting that small artery disease was the most likely cause of PMM. Another large study found no particular etiology in arm or leg monoparesis subgroups [1].

A study of seven patients with superficial infarctions and monoparesis (leg, 2; arm, 5) revealed cardioembolism as the cause in two cases; however, neither had significant carotid stenosis [10]. Another report showed that, out of 14 patients, arm PMM or monoparesis due to IS resulted from carotid pathology and cardioembolism in six and four cases, respectively [13]. In studies that found a correlation between upper limb monoparesis and parietal lobe infarction, all six patients had infarctions in the vascular border zone of angular gyrus white matter; five of these six cases had carotid stenosis, suggesting a hemodynamic mechanism [21]. Others showed that cardiac sources were the most frequent etiologies: cardioembolism/cardiac source in 46%, large artery atherosclerosis/carotid stenosis in 34%, MCA branch thrombosis in 17%, and cryptogenic in 3% [15]. In a study of 11 patients with distal arm PMM due to cortical infarctions, three had a patent foramen ovale, three had carotid stenosis, and two had cardiac arrhythmias; one of the latter also had MCA stenosis [28]. A study of six patients with isolated upper limb weakness due to cortical infarction reported cardioembolism, large-artery atherosclerosis of the MCA, and undetermined etiology in one, two, and three cases, respectively [22]. In a report of 29 precentral knob infarctions (22 of 29 patients showed isolated hand paresis, and 10 of 29 patients showed sensory disturbance), ten patients had carotid stenosis, whereas four were found to have a cardiac embolic source [35]. These results indicated that precentral knob area infarctions were frequently associated with atherosclerotic changes in the carotid artery and suggested an arterioarterial thrombembolic stroke mechanism. Yet in another study of seven patients with arm PMM due to precentral gyrus infarctions, the mechanism of stroke was determined as large artery atherothrombosis (internal carotid artery stenosis, 2 cases), cardiogenic embolism (2 cases), secondary to hypoperfusion (1 case), and undetermined in (2 cases) [39].A recent large study including patients with hand PMM due to knob infarctions had 15 patients with hand palsy, two of whom had sensory disturbances. In that study, stroke etiologies were large artery atherosclerosis in three patients, cardioembolism in two patients, small artery disease in five patients, and unknown in five patients [40]. Kim proposed that the motor cortex region representing ulnar fingers might be a border zone between large arteries. The author suggested that predominant involvement of ulnar fingers due to cortical infarctions was closely associated with severe proximal arterial stenosis or occlusion of the MCA or internal carotid artery and that predominant involvement of radial-sided fingers was often related to emboligenic stroke [29]. Other case reports showed carotid stenosis [34, 36, 45] or cardiogenic embolism [31, 37] as underlying stroke etiologies in PMM patients. One case report showed that hypotension might either cause or contribute to the precentral knob infarction in a patient with hand PMM [26]. The first reported isolated shoulder palsy due to a small cortical infarction with severe carotid stenosis was reported as a border-zone infarction [50]. In fact, proximal arm weakness has traditionally been considered a manifestation of border-zone infarction. However, subsequent isolated shoulder palsy cases due to cardiogenic embolism [51, 54] and atherothrombosis [52, 53] indicated that the mechanism of stroke in arm PMM due to cortical infarction might include both cardioembolic and carotid sources, which underlies the difficulty in determining the more frequent etiologies in these clinical entities. Rare etiologies reported by several groups include aortogenic embolism [41] and cerebral angitis [42] presenting as precentral knob infarctions and PMM.

One study including six patients showed that the mechanism of leg PMM due to small ACA cortical infarctions was cardiac embolism in three cases; the etiology was unknown in the remaining three patients [64]. Other reports also showed carotid stenosis or carotid emboli [8, 59, 63, 73], ACA stenosis [67], and cardiogenic embolism [59, 68] as causes of leg PMM due to ACA cortical infarctions. One study showed that the majority of leg PMM cases were correlated with occlusion or severe stenosis of the internal carotid artery [75]. Additionally, a large study showed that watershed infarctions were found only in patients with leg PMM but not in those with arm PMM [2].

Centrum semiovale infarcts with upper limb monoparesis were reported to result from small vessel disease [76]. Furthermore, lacunar infarctions of the internal capsule were reported to cause PMM [59, 86, 87]. Small artery disease might be the most frequent mechanism for brainstem infarctions [94, 95]; however, it can also result from vertebral artery occlusion [101].

DIFFERENTIAL DIAGNOSIS

As described above, either the radial or ulnar side can be predominantly affected in cortical hand PMM, raising the possibility of misdiagnosis of these patients with cervical disc disease or radial or ulnar neuropathy [5]. It is critical to include cerebral lesions in the differential diagnosis of monoparesis. Following are the important points for a successful diagnosis of PMM due to IS [4]: (1) *Anamnesis.* Typically, mononeuropathy, such as radial nerve palsy, has a preceding compression event; thus, sudden or acute onset of weakness in one hand without history of a compression event might be an indicator of PMM due to IS. (2) *Risk factors for cerebrovascular disease.* The presence of hypertension, diabetes mellitus, atrial fibrillation, and/or hyperlipidemia is also important. (3) *Pyramidal signs.* DTRs are often decreased and no pathological reflexes are present with peripheral nervous system damage. Therefore, DTRs and muscle tone are important tools to distinguish central and peripheral weakness. However, PMM due to IS does not always show active DTRs in the affected limbs. In fact, there is a relatively low frequency of active DTRs in both arm and leg PMM cases resulting from IS [13, 68], and the absence of pyramidal signs should not exclude PMM due to IS. (4) *The distribution of weakness.* Patients with PMM due to IS show frequent distal-dominant weakness, and some have radial or ulnar finger-predominant weakness. In these patients, the affected muscles are not completely the same as the peripheral nerve distribution, and careful neurological evaluation is necessary. (5) *Neuroimaging.* DWI is the most useful tool for diagnosis, as CT and conventional MRI cannot distinguish between acute and chronic lesions. In fact, 11 of 28 patients with PMM due to IS had no visible lesions by CT or conventional MRI [11], and all eight patients with PMM in one study showed no visible lesions on repeat CT scans [87]. In a study of distal arm PMM patients, 77% of the 35 patients demonstrated no causative lesions on their first CT scans [15]. Without the use of DWI, small cortical infarctions as well as acute infarctions might be overlooked or misidentified. The recommended MRI sequences include DWI for acute ischemia, fluid-attenuated inversion recovery, T2*-weighted gradient-echo images for bleeding, and MR angiography.

PROGNOSIS

A large study in 2005 showed that 41% of PMM patients returned to their former activities and that 41% of patients required some assistance, indicating that outcomes of PMM are generally favorable [1]; however, isolated facial paresis and hemorrhagic stroke were observed in some patients. Therefore, the exact outcomes of PMM due to IS remain unclear.

Some case reports on cortical infarctions presenting as arm PMM described improvements [14, 24, 42, 56]. While most studies did not evaluate the prognosis of PMM due to IS [8, 11, 12, 25, 27, 32, 36, 38, 43, 44, 50], others determined good or complete recovery in patients with PMM due to IS [10, 15, 22, 23, 26, 28-31, 33-35, 37, 41, 45, 46, 51-53, 106]. For example, one study determined that 14 of the 35 patients with arm PMM due to cortical infarctions had complete clinical recovery, albeit residual mild weakness was reported in 21 patients [15]. Yet another report showed that all patients with PMM had complete recovery within 2–4 weeks [10]. Furthermore, several case studies showed a high frequency of complete recovery [22, 28, 35]. Finally, patients exhibited good recovery despite the presence of severe finger weakness at initial examination [23, 29, 34, 37, 45].

The rate of recurrent stroke in PMM due to IS has not been established; however, several studies examined outcomes in patients with PMM of the arm. Specifically, one report on arm PMM due to cortical infarction showed a 14% rate of recurrent stroke (mean follow-up, 1.7 years) [15], whereas another study of seven patients with arm PMM due to cortical infarctions with a mean follow-up of 14 months showed two patients had stroke recurrence. The second study also reported the death of one patient; the remaining four cases had excellent recovery with no recurrence [39]. A recent study of 15 patients with knob infarctions, of which 13 had PMM and 2 had monoparesis with sensory disturbances, had a mean follow-up period of 29.8 months [40]. While none of the patients had recurrence, one patient died, and two patients had bad recovery; the hand function was good and fair in six and eight cases, respectively [40]. In some studies, patients with PMM underwent carotid endarterectomy for carotid stenosis [15, 34, 36, 45]. Finally, patients with isolated shoulder PMM due to cortical infarction were reported to have good recovery [51-53]; however, in one study, one patient died on day ten [54].

In a study of cortical ACA infarction cases presenting with lower limb PMM, two of the three patients with proximal weakness had good clinical recovery, whereas the only patient with distal weakness had poor recovery [59]. A separate report showed that five of the six patients with distal leg PMM had partial or complete clinical recovery [64]. Other studies reported recoveries ranging from complete [62, 66] to good [70, 72] or improved [60, 61, 63, 65, 67, 69, 74]. One patient with a subcortical infarction had a good prognosis [77]. Prognoses have not been described for patients with subcortical, centrum semiovale, or corona radiata infarctions presenting as PMM. Patients with pontine infarctions presenting as PMM can have good recoveries; three cases with leg PMM due to rostral lateral pontine infarctions showed good functional recovery [95], and one case with brachial PMM due to a pontine infarction showed a complete recovery [93]. Furthermore, gradual improvements or complete recovery were observed in other cases with brachial PMM due to pontine infarctions or with leg PMM due to medullary infarctions [94, 101, 104].

In summary, previous studies have indicated a good prognosis for patients with arm PMM due to cortical infarctions, especially those in the precentral knob. Due to their rarity, the outcomes of patients with PMM due to infarctions in other regions are difficult to discuss; however, most reported cases had good clinical recoveries. Ultimately, both short- and long-term outcomes depend on risk factors and stroke mechanisms [40].

FURTHER INVESTIGATION

The history of studies about PMM due to IS provides anatomical evidence regarding motor cortical representations in human cortex and differences between motor fiber pathways of the arm and leg; however, some anatomical questions remain debatable. Tractography studies were not always consistent with clinical experiences in patients with PMM due to IS. Although some individual variation exists, it is unclear whether motor somatotopy of the arm and leg clearly exists or breaks down in the deeper regions, especially the brainstem. Moreover, whether individual fingers are represented in a separate cortical territory with overlap should be reexamined as certain case reports indicated that isolated thumb, index finger, or ring finger palsies could be the sole manifestation of cortical infarction [4]. Finally, individual motor somatotopy of the toe was not well understood until now.

Despite its importance, PMM due to IS remains underestimated. In recent years, it has been increasingly reported by neurologists who recognize and understand this atypical stroke syndrome. However, orthopedists and general clinicians are mostly unaware of this condition, and neurologists should alert physicians to its presence. Two large studies from 2005 were hospital-based research and did not use MRI, especially DWI, in all patients. Moreover, the percentage of patients who had PMM in those studies was probably smaller than the general population as patients with very minor motor deficits are less likely to be admitted to a hospital [2] or are misdiagnosed with nonvascular causes of weakness. Additionally, the assumption that lacunar etiology is not likely a cause of PMM due to IS should be reexamined [2]. The exact frequency of PMM due to IS, its clinical features such as paresis pattern, DTR, and the presence of Babinski sign, and prognosis, especially the recurrent stroke rate, should be examined in a large study by a national survey, using DWI for all cases.

REFERENCES

[1]   Maeder-Ingvar M, van Melle G, Bogousslavsky J. Pure monoparesis: a particular stroke subgroup? *Arch Neurol.* 2005; 62:1221-1224.

[2]   Paciaroni M, Caso V, Milia P, Venti M, Silvestrelli G, Palmerini F, Nardi K, Micheli S, Agnelli G. Isolated monoparesis following stroke. *J Neurol Neurosurg Psychiatry.* 2005;76:805-807.

[3]   Isayev Y, Castaldo J, Rae-Grant A, Barbour P. Pure monoparesis: what makes it different. *Arch Neurol.* 2006;63:786.

[4]   Hiraga A. Pure motor monoparesis due to ischemic stroke. *Neurologist.* 2011;17:301-308.

[5]   Edlow JA, Selim MH. Atypical presentations of acute cerebrovascular syndromes. *Lancet Neurol.* 2011;10:550-560.

[6]   Lhermitte J. Semiological value of disorders of sensitivity to raduculaire available in lesions del'encephale. *Sem Med.* 1909;24:277.

[7]   Ashizawa T, Rolak LA, Hines M. Spastic pure motor monoparesis. *Ann Neurol.* 1986;20:638-641.

[8]   Weisberg LA. Small superficial cerebrovascular lesions: clinical and computed tomographic correlations. *Comput Radiol.* 1984;8:151-156.

[9]   Bogousslavsky J, Van Melle G, Regli F. Middle cerebral artery pial territory infarcts: a study of the Lausanne Stroke Registry. *Ann Neurol.* 1989;25:555-560.

[10]   Boiten J, Lodder J. Isolated monoparesis is usually caused by superficial infarction. *Cerebrovasc Dis.* 1991;1:337-340.

[11]   Melo TP, Bogousslavsky J, van Melle G, Regli F. Pure motor stroke: a reappraisal. *Neurology.* 1992;42:789–795.

[12]   Mohr JP, Foulkes MA, Polis AT, Hier DB, Kase CS, Price TR, Tatemichi TK, Wolf PA. Infarct topography and hemiparesis profiles with cerebral convexity infarction: the Stroke Data Bank. *J Neurol Neurosurg Psychiatry.* 1993;56:344-351.

[13]   Iqbal J, Bruno A, Berger M. Stroke causing pure brachial monoparesis. *J Stroke Cerebrovasc Dis.* 1995;5:88-90.

[14]   Yousry TA, Schmid UD, Alkadhi H, Schmidt D, Peraud A, Buettner A, Winkler P. Localization of the motor hand area to a knob on the precentral gyrus: a new landmark. *Brain.* 1997;120:141-157.

[15]   Castaldo J, Rodgers J, Rae-Grant A, Barbour P, Jenny D. Diagnosis and neuroimaging of acute stroke producing distal arm monoparesis. *J Stroke Cerebrovasc Dis.* 2003;12:253-258.

[16]   de Freitas GR, Devuyst G, van Melle G, Bogousslavsky J. Motor strokes sparing the leg: different lesions and causes. *Arch Neurol.* 2000;57:513-518.

[17]   Fraix V, Besson G, Hommel M, Perret J. Brachiofacial pure motor stroke. *Cerebrovasc Dis.* 2001;12:34-38.

[18]   Yoneda Y, Mori E, Tabuchi M, Yamadori A. Pure motor monoparesis due to intracerebral hemorrhage. *Stroke.* 1993;24:142-143.

[19]   Luo JJ, Azizi AS. Hematoma causing cortical hand. *Neurology.* 2002;59:E12.

[20]   Hiraga A, Kamitsukasa I. Isolated foot drop due to subcortical hemorrhage. *J Neurol.* 2010;257:1741-1742.

[21]   Timsit S, Logak M, Manaï R, Rancurel G.. Evolving isolated hand palsy: a parietal lobe syndrome associated with carotid artery disease. *Brain.* 1997;120:2251-2257.

[22]   Chen PL, Hsu HY, Wang PY. Isolated hand weakness in cortical infarctions. *J Formos Med Assoc.* 2006;105:861-865.

[23] Lee PH, Han SW, Heo JH. Isolated weakness of the fingers in cortical infarction. *Neurology.* 1998;50:823-824.

[24] Tei H. Monoparesis of the right hand following a localized infarct in the left precentral knob. *Neuroradiology.* 1999;41:269-270.

[25] Phan TG, Evans BA, Huston J. Pseudoulnar palsy from a small infarct of the precentral knob. *Neurology.* 2000;54:2185.

[26] Altieri M, Di Piero V, Bastianello S, Luigi Lenzi G.. Hand weakness from a precentral gyrus infarct with intermittent hypotension. *Neurology.* 2001;56:1748.

[27] Back T, Mrowka M. Infarction of the "hand knob" area. *Neurology.* 2001;57:1143.

[28] Gass A, Szabo K, Behrens S, Rossmanith C, Hennerici M. A diffusion-weighted MRI study of acute ischemic distal arm paresis. *Neurology.* 2001;57:1589-1594.

[29] Kim JS. Predominant involvement of a particular group of fingers due to small, cortical infarction. *Neurology.* 2001;56:1677–1682.

[30] Takahashi N, Kawamura M, Araki S. Isolated hand palsy due to cortical infarction: localization of the motor hand area. *Neurology.* 2002;58:1412–1414.

[31] Kim JS, Chung JP, Ha SW. Isolated weakness of index finger due to small cortical infarction. *Neurology.* 2002;58:985.

[32] Uribe Roca C, Gatto EM, Micheli F. Isolated weakness of index finger due to small cortical infarction. *Neurology.* 2002;59:2010-2012.

[33] Kobayashi M, Sonoo M, Shimizu T. Pure motor stroke with major involvement of the index finger. *J Neurol Neurosurg Psychiatry.* 2004;75:507-508.

[34] Sudo K, Kishimoto R, Tajima Y, Matsumoto A, Tashiro K. A paralysed thumb. *Lancet.* 2004; 363:1364.

[35] Peters N, Müller-Schunk S, Freilinger T, Düring M, Pfefferkorn T, Dichgans M. Ischemic stroke of the cortical "hand knob" area: stroke mechanisms and prognosis. *J Neurol.* 2009;256:1146-1151.

[36] Hall J, Flint AC. Neurological picture. "Hand knob" infarction. *J Neurol Neurosurg Psychiatry.* 2008;79:406.

[37] Granziera C, Kuntzer T, Vingerhoets F, Cereda C. Small cortical stroke in the "hand knob" mimics anterior interosseous syndrome. *J Neurol.* 2008;255:1423-1424.

[38] Cornely C, Muhl C, Himstedt J, Isenmann S. Isolated ring finger palsy due to cortical infarction of the primary motor cortex hand area. *Clin Neurophysiol.* 2009;120:e42.

[39] Pikula A, Stefanidou M, Romero JR, Kase CS. Pure motor upper limb weakness and infarction in the precentral gyrus: mechanisms of stroke. *J Vasc Interv Neurol.* 2011;4:10-13.

[40] Alstadhaug KB, Sjulstad A. Isolated hand paresis: a case series. *Cerebrovasc Dis Extra. 2013*;3:65-73.

[41] Kawabata Y, Miyaji Y, Joki H, Seki S, Mori K, Kamide T, Tamase A, Nomura M, Kitamura Y, Tanaka F. Isolated index finger palsy due to cortical infarction. *J Stroke Cerebrovasc Dis.* 2014;23:e475-476.

[42] Jusufovic M, Lygren A, Aamodt AH, Nedregaard B, Kerty E. Pseudoperipheral palsy: a case of subcortical infarction imitating peripheral neuropathy. *BMC Neurol.* 2015 25;15:151.

[43] Schieber MH. Somatotopic gradients in the distributed organization of the human primary motor cortex hand area: evidence from small infarcts. *Exp Brain Res.* 1999;128:139-148.

[44] Celebisoy M, Ozdemirkiran T, Tokucoglu F, Kaplangi DN, Arici S. Isolated hand palsy due to cortical infarction: localization of the motor hand area. *Neurologist.* 2007;13:376-379.

[45] Hochman MS, DePrima SJ, Leon BJ. Early diagnosis by diffusion-weighted MRI of pure motor stroke limited to finger weakness. *J Neuroimaging.* 1998; 8:179-181.

[46] Rankin EM, Rayessa R, Keir SL. Pseudoperipheral palsy due to cortical infarction. *Age Ageing.* 2009; 38:623-624.

[47] Wada K, Kimura K, Minematsu K, Uchino M, Yamaguchi T. Spotty cortical enhancement detected by magnetic resonance imaging: a sign of embolic transient ischemic attack and stroke? *J Stroke Cerebrovasc Dis.* 2001;10:19-22.

[48] Penfield W, Boldrey E. Somatic motor and sensory representation in the cerebral cortex of man as studied by electrical stimulation. *Brain.* 1937;60:389-443

[49] Terao Y, Hayashi H, Kanda T, Tanabe H. Discrete cortical infarction with prominent impairment of thumb flexion. *Stroke.* 1993;24:2118-2120.

[50] Komatsu K, Fukutake T, Hattori T. Isolated shoulder paresis caused by a small cortical infarction. *Neurology.* 2003;61:1457.

[51] Nah HU, Park HK, Kang DW. Isolated shoulder weakness due to a small cortical infarction. *J Clin Neurol.* 2006;2:209-211.

[52] Tsuda H, Kubota Y, Tanaka K, Kishida S. Isolated shoulder palsy due to a cortical infarction. *Intern Med.* 2011;50:947.

[53] Tsuda H, Shinozaki Y, Tanaka K, Miura Y, Kishida S, Karasawa K. Isolated shoulder palsy due to infarction of the cortical branch of the middle cerebral artery. *Intern Med.* 2012;51:2217-2219.

[54] Kawasaki A, Suzuki K, Takekawa H, Kokubun N, Yamamoto M, Asakawa Y, Okamura M, Hirata K. Isolated shoulder palsy due to cortical infarction: a case report and literature review of clinicoradiological correlations. *J Stroke Cerebrovasc Dis.* 2013; 22:e687-690.

[55] Hiraga A, Ito S, Hayakawa S, Misawa S, Hattori T. Focal shoulder seizures due to an acute small cortical infarction located medial to precentral knob. *J Neurol.* 2006;253:963-964.

[56] Uncini A, Caporale CM, Caulo M, Ferretti A, Tartaro A, Ranieri F, Di Lazzaro V. Isolated shoulder palsy due to cortical infarction: localisation and electrophysiological correlates of recovery. *J Neurol Neurosurg Psychiatry.* 2007;78:100-102.

[57] Critcheley M. The anterior cerebral artery and its syndrome. *Brain.* 1930:52:120-165.

[58] Bogousslavsky J, Regli F. Anterior cerebral artery territory infarction in the Lausanne Stroke Registry: clinical and etiologic patterns. *Arch Neurol.* 1990;47:144-150.

[59] Schneider R, Gautier JC. Leg weakness due to stroke: site of lesions, weakness patterns and causes. *Brain.* 1994;117:347-354.

[60] Kohno Y, Ohkoshi N, Shoji S. Pure motor monoparesis of a lower limb due to a small infarction in the contralateral motor cortex. *Clin Imaging.* 1999;23:149-151.

[61] Han IB, Ahn JY, Chung YS, Chung SS. Isolated distal leg weakness due to a small cerebral infarction masquerading as a spinal lesion. *J Korean Neurosurg Soc.* 2007;41:182-185.

[62] Ku BD, Lee EJ, Kim H. Cerebral infarction producing sudden isolated foot drop. *J Clin Neurol.* 2007;3:67-69.

[63] Noda K, Tani M, Fukae J, Fujishima K, Hattori N, Okuma Y. Isolated proximal leg paresis due to a small cortical infarction. *Intern Med.* 2010;49:1633-1636.

[64] Alonso A, Gass A, Griebe M, Kern R, Rossmanith C, Hennerici MG, Szabo K. Isolated ischaemic lesions in the foot motor area mimic peripheral lower-limb palsy. *J Neurol Neurosurg Psychiatry.* 2010;81:822-823.

[65] Kim KW, Park JS, Koh EJ, Lee JM. Cerebral infarction presenting with unilateral isolated foot drop. *J Korean Neurosurg Soc.* 2014; 56:254-256.

[66] Ricarte IF, Figueiredo MM, Fukuda TG, Pedroso JL, Silva GS. Acute foot drop syndrome mimicking peroneal nerve injury: an atypical presentation of ischemic stroke. *J Stroke Cerebrovasc Dis.* 2014;23:1229-1231.

[67] Kim JY, Kim do K, Yoon SH. Isolated painless foot drop due to cerebral infarction mimicking lumbar radiculopathy: a case report. *Korean J Spine.* 2015;12:210-212.

[68] Hiraga A, Uzawa A, Tanaka S, Ogawara K, Kamitsukasa I. Pure monoparesis of the leg due to cerebral infarctions: a diffusion-weighted imaging study. *J Clin Neurosci.* 2009; 16:1414-1416.

[69] Nagaratnam N, Beh P, Ooi EA. Pure motor monoparesis. *Br J Clin Pract.* 1990;44:796-797.

[70] Nagaratnam N, Davies D, Chen E. Clinical effects of anterior cerebral artery infarction. *J Stroke Cerebrovasc Dis.* 1998;7:391-397.

[71] Kumral E, Bayulkem G, Evyapan D, Yunten N. Spectrum of anterior cerebral artery territory infarction: clinical and MRI findings. *Eur J Neurol.* 2002;9:615-624.

[72] Cattaneo L, Cucurachi L, Pavesi G. Neurological picture. Isolated toe paralysis caused by a small cortical infarction. *J Neurol Neurosurg Psychiatry.* 2009;80:1142.

[73] Soga K, Irioka T, Higashi M, Mizusawa H. Stroke presenting with monoparesis in the lower limb. *Intern Med.* 2012;51:819-820.

[74] Park KM, Kim SE, Shin KJ, Park J, Kim SE, Kim HC, Mun CW, Ha SY. Isolated foot drop in acute infarction of the supplementary motor area. *Clin Neurol Neurosurg.* 2013; 115:2240-2242.

[75] Yanagihara T, Sundt TM Jr, Piepgras DG. Weakness of the lower extremity in carotid occlusive disease. *Arch Neurol.* 1988;45:297-301.

[76] Bogousslavsky J, Regli F. Centrum ovale infarcts: subcortical infarction in the superficial territory of the middle cerebral artery. *Neurology.* 1992;42:1992-1998.

[77] Hochman MS, DePrima SJ, Leon BJ. Diffusion-weighted MRI diagnosis of pure motor stroke limited to primarily distal leg weakness. *J Neuroimaging.* 2000;10:118-120.

[78] Chamorro A, Sacco RL, Mohr JP, Foulkes MA, Kase CS, Tatemichi TK, Wolf PA, Price TR, Hier DB. Clinical-computed tomographic

correlations of lacunar infarction in the Stroke Data Bank. *Stroke.* 1991;22:175-181.

[79] Song YM. Somatotopic organization of motor fibers in the corona radiata in monoparetic patients with small subcortical infarct. *Stroke.* 2007;38:2353-2355.

[80] Tohgi H, Takahashi S, Takahashi H, Tamura K, Yonezawa H. The side and somatotopical location of single small infarcts in the corona radiata and pontine base in relation to contralateral limb paresis and dysarthria. *Eur Neurol.* 1996;36:338-342.

[81] Kim JS, Pope A. Somatotopically located motor fibers in corona radiata: evidence from subcortical small infarcts. *Neurology.* 2005;64:1438-1440.

[82] Lee DH, Lee DW, Han BS. Topographic organization of motor fibre tracts in the human brain: findings in multiple locations using magnetic resonance diffusion tensor tractography. *Eur Radiol.* 2015 [Epub ahead of print]

[83] Seo JP, Chang PH, Jang SH. Anatomical location of the corticospinal tract according to somatotopies in the centrum semiovale. *Neurosci Lett.* 2012;523:111-114.

[84] Goto H, Tanaka T, Momozaki N. Subcortical infarction causes pure motor isolated finger palsy. *Intern Med.* 2013;52:1283-1284.

[85] Rascol A, Clanet M, Manelfe C, Guiraud B, Bonafe A. Pure motor hemiplegia: CT study of 30 cases. *Stroke.* 1982;13:11-17.

[86] Donnan GA, Tress BM, Bladin PF. A prospective study of lacunar infarction using computerized tomography. *Neurology.* 1982;32:49-56.

[87] Arboix A, Marti-Vilalta JL, Garcia JH. Clinical study of 227 patients with lacunar infarcts. *Stroke.* 1990; 21:842-847.

[88] Maeda K, Kawai H, Nakamura H, Idehara R, Ogawa N, Sanada M, Yasuda H. Internal capsule infarction showing leg monoparesis. *Intern Med.* 2007;46:1273.

[89] Bertrand G, Blundell J, Musella R. Electrical exploration of the internal capsule and neighbouring structures during stereotaxic procedures. *J Neurosurg.* 1965;22:333-343.

[90] Holodny AI, Gor DM, Watts R, Gutin PH, Ulug AM.. Diffusion-tensor MR tractography of somatotopic organization of corticospinal tracts in the internal capsule: initial anatomic results in contradistinction to prior reports. *Radiology.* 2005;234:649-653.

[91] Ino T, Nakai R, Azuma T, Yamamoto T, Tsutsumi S, Fukuyama H. Somatotopy of corticospinal tract in the internal capsule shown by

functional MRI and diffusion Tensor images. *Neuroreport.* 2007;18:665-668.

[92] Lee DH, Kwon YH, Hwang YT, Kim JH, Park JW. Somatotopic location of corticospinal tracts in the internal capsule with MR tractography. *Eur Neurol.* 2012;67:69-73.

[93] Iwasaki Y, Kinoshita M, Ikeda K, Takamiya K, Shiojima T. A case of pure motor monoparesis due to pontine infarction. *Int J Neurosci.* 1990;55:157-159.

[94] Song IU, Kim JS, Kim YI, Lee KS. Brachial monoparesis as an isolated manifestation of a paramedian pontine ischemic lesion. *Clin Neurol Neurosurg.* 2007;109:376-378.

[95] Kataoka S, Miaki M, Saiki M, Saiki S, Yamaya Y, Hori A, Hirose G.. Rostral lateral pontine infarction: neurological/topographical correlations. *Neurology.* 2003;61:114-117.

[96] Kataoka S, Hori A, Shirakawa T, Hirose G. Paramedian pontine infarction: neurological/topographical correlation. *Stroke.* 1997;28:809-815.

[97] Schmahmann JD, Ko R, MacMore J. The human basis pontis: motor syndromes and topographic organization. *Brain.* 2004;12:1269-1291.

[98] Marx JJ, Iannetti GD, Thömke F, Fitzek S, Urban PP, Stoeter P, Cruccu G, Dieterich M, Hopf HC. Somatotopic organization of the corticospinal tract in the human brainstem: a MRI-based mapping analysis. *Ann Neurol.* 2005;57:824-831.

[99] Hong JH, Son SM, Jang SH. Somatotopic location of corticospinal tract at pons in human brain: a diffusion tensor tractography study. *Neuroimage.* 2010;51:952-955.

[100] Kwon HG, Hong JH, Lee MY, Kwon YH, Jang SH. Somatotopic arrangement of the corticospinal tract at the medullary pyramid in the human brain. *Eur Neurol.* 2011;65:46-49.

[101] Tsuda H, Tanaka K, Kishida S. Pure motor monoparesis in the leg due to a lateral medullary infarction. *Case Rep Med.* 2012;2012:758482.

[102] Kim JS, Kim HG, Chung CS. Medial medullary syndrome: report of 18 new patients and a review of the literature. *Stroke.* 1995;26:1548-1552.

[103] Kim JS, Han YS. Medial medullary infarction: clinical, imaging, and outcome study in 86 consecutive patients. *Stroke.* 2009;40:3221-3225.

[104] Fumoto N, Tomimoto H, Inoue H, Fukuyama H, Takahashi R. Medial medullary infarction with monoparesis of lower extremity: a case report. *Neurol Med* (Tokyo). 2008;68:92-94 (Japanese).

[105] Chimowitz MI, Furlan AJ, Sila CA, Paranandi L, Beck GJ. Etiology of motor or sensory stroke: a prospective study of the predictive value of clinical and radiological features. *Ann Neurol.* 1991;30:519-525.

[106] Lüders HO. Isolated hand palsy due to cortical infarction: Localization of the motor hand area. *Neurology.* 2003;60:1054-1055.

[107] Nelson RF, Pullicino P, Kendall BE, Marshall J. Computed tomography in patients presenting with lacunar syndromes. *Stroke.* 1980;11:256-261.

In: Horizons in Neuroscience Research. Vol. 25   ISBN: 978-1-63485-286-9
Editors: A. Costa and E. Villalba   © 2016 Nova Science Publishers, Inc.

Chapter 9

STROKE ASSOCIATED WITH NEPHROTIC SYNDROME - ACUTE ISCHEMIC STROKE, CEREBRAL VENOUS SINUS THROMBOSIS, AND INTRACEREBRAL HEMORRHAGE

Masaru Kuriyama, MD, PhD

Department of Neurology, Brain Attack Center Ota Memorial Hospital,
Fukuyama, Hiroshima, Japan

ABSTRACT

Nephrotic syndrome is defined by the presence of heavy proteinuria, hypoalbuminemia, hyperlipidemia and peripheral edema. Thrombotic diseases are also frequently observed. Arterial and venous thromboses are potential complications of nephrotic syndrome. Arterial thromboses are less frequent than venous thromboses and the most common locations are femoral arteries, although other arteries may be involved. Stroke associated with nephrotic syndrome has been reported in a number of case reports, but it is rare and its clinical details remain obscure. We identified ten cases of acute ischemic stroke, two cases of cerebral venous sinus thrombosis, and four cases of intracerebral hemorrhage by screening the hospitalized patients enrolled in our stroke unit. The incidence of acute ischemic stroke associated with nephrotic syndrome was 0.09% of all kinds of stroke and 0.12% of acute ischemic stroke, and was fivefold that of cerebral venous sinus thrombosis. The subtypes of stroke identified in our screening were large-artery atherosclerosis (six instances), small-vessel occlusion (three instances), and cardioembolism (one instance). The results of a retrospective cohort study showed that acute ischemic stroke-associated nephrotic syndrome is strongly associated with atherosclerosis of the cerebral artery, especially in the anterior circulation. Patients also exhibited atherosclerosis of the internal carotid and lower extremity arteries. The causal disorder of nephrotic syndrome was diabetic nephropathy in eight of these cases. Twenty-five patients with cerebral venous sinus thrombosis were admitted to our hospital during the same period. Nephrotic syndrome has been reported to be responsible for 0.6-1.0% of all cases of cerebral venous sinus thrombosis; however, our study showed that this percentage may be higher. Nephrotic syndrome essentially induced a hypercoagulable state. However, four cases of intracerebral hemorrhage associated with nephrotic syndrome were identified by the screening. These exhibited the presence of strong risk factors for hemorrhage, including hypertension, vascular vulnerability with atherosclerosis or amyloidosis, and change of bloodstream after operation, suggesting that they overcame the risk for thrombophilia. The diseases associated with nephrotic syndrome were diabetic nephropathy and amyloidosis in three cases and one case, respectively. Nephrotic syndrome tends to be associated with a risk for arterial or venous thrombosis. In addition, we must pay attention to intracerebral hemorrhage associated with nephrotic syndrome in cases of stroke.

1. INTRODUCTION

Nephrotic syndrome (NS) is defined by the presence of heavy proteinuria (protein excretion greater than 3.5 g/24 hours), hypoalbuminemia (less than 3.0 g/dL), hyperlipidemia, and peripheral edema [1] (Figure 1). Thrombotic diseases are also frequently observed. Arterial and venous thromboses are potential complications of NS. Arterial thromboses are less frequent than venous thromboses, and the most common locations are femoral arteries, although other arteries may be involved [1-3]. Acute ischemic stroke (AIS) associated with NS has been reported in several case reports [4-10], but this is relatively rare.

Cerebral venous sinus thrombosis (CVT) is less common than most other types of stroke, but is recognized with increasing frequency due to the widespread use of magnetic resonance imaging (MRI) and rising clinical awareness. The clinical severity varies, and severe sequelae can occur when treatment is delayed. In some cases, death may occur from brain herniation due to increased intracranial pressure; therefore, quick and accurate diagnosis is required.

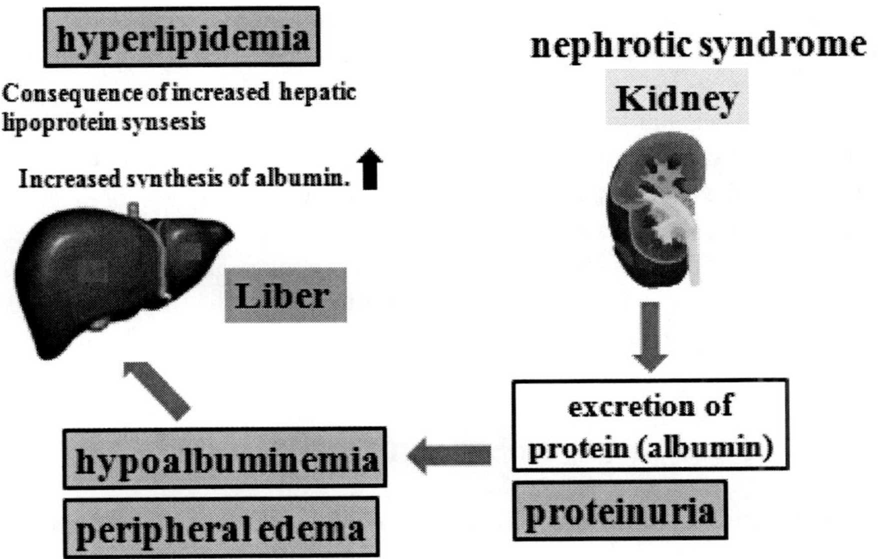

Figure 1. Clinical symptoms in nephrotic syndrome.

As CVT is also caused by various underlying diseases, it is important to analyze the causative disorders [2, 11, 12], as well as the immediate treatment for thrombosis. In the past, CVT was attributed to infections of the face and otomastoid areas, but since the introduction of antibiotics, it is more often related to pregnancy, puerperium, neoplasms, dehydration, genetic or acquired thrombophilia, and oral contraceptives [2, 11-13]. In Western countries in particular, CVT is increased by the hypercoagulability induced by oral contraceptive use [14-16]. NS is one of the risk factors for CVT. With recent increases in the number of patients with NS, it is likely that the number of patients with CVT caused by NS will increase, especially in Japan.

Hypercoagulability contributes to the predisposition to thromboembolism in NS. The low molecular weight coagulation factors (Factor IX, XI), antithrombin III, plasminogen, and free protein S excreted into the urine by the breakdown of the permselectivity barrier of the glomerular capillary wall have been reported, along with inverse increases in high molecular weight coagulation factors (Factor V, VII, VIII and X), fibrinogen, alpha 2-antiplasmin, alpha-2-macroglobulin and platelet aggregation [16-23]. These changes have been attributed to a hypercoagulable state in which an imbalance between procoagulant/prothrombotic factors and anticoagulant/antithrombotic factors promote thromboses in veins and arteries. Hypoalbuminemia, secondary volume depletion, and hyperlipidemia also could act as risk factors for thromboembolism or atherosclerosis. NS may independently predispose individuals to arterial and venous thromboembolism, but the exact mechanism of the hypercoagulability is as yet not fully understood [1-4, 16-23].

Hypercoagulability due to NS involves the pathogenesis of AIS or CVT. When we retrospectively screened stroke patients with NS from hospitalized patients enrolled in our stroke registry, it was found not only in those patients with AIS and CVT, but also, unexpectedly, in some patients with intracerebral hemorrhage (ICH) [24-26]. These patients had strong risk factors for ICH, suggesting that they had overcome the risk for thrombophilia. Attention should be paid to NS as a risk factor for stroke including AIS, CVT, and also ICH.

2. SCREENING FOR STROKE PATIENTS WITH NS

We retrospectively analyzed 11,161 cases of hospitalized patients enrolled in our stroke registry during a nine-year period from April 2004 to August 2013. The number of patients with serum albumin lower than 3.0 g/dl was 196 (1.9%) out of 10,057 patients examined, and that of patients with serum cholesterol higher than 250 mg/dl was 1,201 (11.3%) of 10,619 patients examined. The patients with albumin below 3.0 g/dl and serum cholesterol above 250 mg/dl at the same time were screened, and only 21 cases fulfilled both levels. From these patients, furthermore, the cases showing heavy proteinuria were selected. The proteinuria was estimated quantitatively by measuring the protein in a urine/24 hours or qualitatively by using a paper kit (Eiken Chemical Co. Japan) (Figure 2). They included two cases showing negative or equivocal (±) results in the protein paper kit test, one case with 1+ (30 mg/dl), two cases with 2+ (100 mg/dl), five cases with 3+ (300 mg/dl), and 11 cases with 4+ (1000 mg/dl). The 16 patients showed urinary protein 3+ or 4+ - i.e., heavy proteinuria – and comprised 10 patients with AIS, four with intracranial hemorrhage, and two with cerebral venous thrombosis (CVT) (Table 1). Nine of these 16 patients showed protein excretion greater than 3.5 g/24 hours. In the other seven patients, the content of protein in urine/24 hours was not measured. However, they all had moderate to severe peripheral edema and hyperlipidemia, and the diagnosis of NS was also strongly suggested in these seven cases.

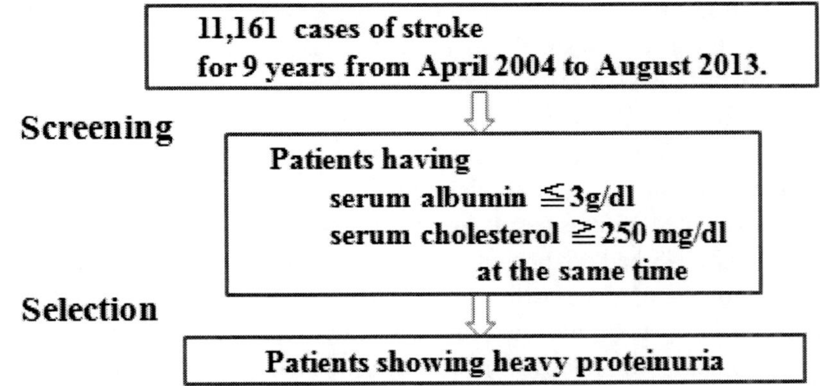

Proteinuria: measured quantitatively by measuring the protein in a urine/24 hours or qualitatively by using paper kit (Eiken Chemical Co. Japan).

Figure 2. The screening and selection of the stroke patients with nephritic syndrome.

Table 1. The patients of acute stroke

	TIA	SVO	LAA	CE	other	ICH	SAH	Total
Patients n.	660	2196	2547	2023	235	2200	733	10619
① TC >250	60 (9.1)	299 (13.6)	373 (14.6)	133 (6.6)	17 (7.2)	242 (11.0)	75 (10.2)	1201 (11.3)
Patients n.	632	2070	2436	1958	202	2055	675	10057
② Alb < 3.0	3 (0.5)	16 (0.8)	44 (1.8)	66 (3.4)	9 (4.5)	48 (2.3)	8 (1.2)	196 (1.9)
Patients n.	263	886	1123	917	99	1226	477	5002
③ massive UP	8 (3.0)	21 (2.4)	40 (3.6)	45 (4.9)	6 (6.1)	118 (9.6)	33 (7.0)	272 (5.4)
① + ② + ③	0	3	6	1	2 (CVT)	4	0	

Abbreviations: TC – total cholesterol; Alb – albumin; UP – urinary protein; TIA – transient ischemic attack; SVO – small-vessel occlusion; LAA – large-artery atherosclerosis; CE – cardioembolism; ICH – intracerebral hemorrhage; SAH – subarachnoid hemorrhage; CVT – cerebral venous sinus thrombosis.

3. ACUTE ISCHEMIC STROKE (AIS) ASSOCIATED WITH NS

1) AIS Patients

The 10 cases – eight males and two females, 64.4 ± 8.6 yrs (Mean ± SD) – were diagnosed as AIS associated with NS. The incidence rate thereof ? was

0.09% of all kinds of stroke (total 11,161 patients) and 0.12% of AIS (total 8,116 patients) in our hospital. Of these ten cases, six were of large-artery atherosclerosis (LAA), three of small-vessel occlusion (lacune, SVO), and one of cardioembolism (CE) by the TOAST criteria. On MRI and MRA, middle cerebral arteries, anterior choroidal artery, and internal carotid artery (occlusion) were involved in six cases, one case and one case, respectively. SVOs were located in the supra-tentorial basal ganglia in all three of the SVO cases (Table 2). NS had not been diagnosed prior to admission in any of the 10 cases. Pathological diagnosis was arrived at through histopathological examination of the renal biopsied samples in two cases; amyloid nephropathy due to primary amyloidosis in one case, and membranoproliferative glomerulonephritis (MPGN) in one case. Eight cases were diagnosed as diabetic nephropathy (DMN) based on the long history of diabetic mellitus (DM) and the exclusion of other pathogenetic diseases, although renal biopsy was not performed (Table 2).

Table 2. The patients of acute ischemic stroke with nephrotic syndrome

Case	Age	Sex	Type	MRI Location of infarction	Alb	TC	UP	day	E	DM Duration (years)	NS type
1	50	F	LAA	R. MCA (BAD)	3	284	4	8.4	++	15	DN s/p
2	53	F	LAA	R. MCA	1.8	707	4	8.6	++	—	AMN
3	59	M	LAA	R. Ant. Choro. A	2.1	336	3	9.5	++	5	DN s/p
4	62	M	LAA	L. MCA	2.8	407	4	6.4	++	30	DN s/p
5	79	M	LAA	R. MCA (R. IC occlusion)	2.8	315	4		++	20	DN s/p
6	70	M	LAA	R. MCA,	1.9	326	3		+	10	DN s/p
7	65	M	SVO	L. Basal Ganglia	2.7	264	4	6.5	++	—	MPGN
8	68	M	SVO	R. Basal Ganglia	3	285	3		++	2	DN s/p
9	69	M	SVO	R. Basal Ganglia	3	299	3		++	4	DN s/p
10	69	M	CE	R. MCA	3	264	4		++	8	DN s/p

Abbreviations: TC – total cholesterol; Alb – albumin; UP – urinary protein, E – edema; SVO – small-vessel occlusion; LAA – large-artery atherosclerosis; CE – cardioembolism; L – left; R – right; MCA – middle cerebral artery; BAD – branched atheromatous disease; Ant – anterior; Choro. –Choroidal-IC – internal carotid; NS – nephrotic syndrome; DN – diabetic nephropathy; MPGN – membranoproliferative glomerulonephritis. AMN – amyloido nephropathy

2) Arterial Stenosis in the Patients

Extent of atherosclerosis was evaluated by the atherosclerosis score of intracranial and extracranial arteries on MRA [27]. A total of 17 vessels – including the common carotid arteries, the exracranial and intracranial portions of the internal carotid arteries and the vertebral arteries, the anterior, middle, and posterior cerebral arteries, and the basilar artery – were evaluated by the cervical and intracranial MRAs. The individual total scores were divided into two scores: an anterior score and a posterior score, related to the anterior circulation vessels (common carotid, internal carotid, anterior cerebral, and middle cerebral arteries) and posterior circulation vessels (vertebral, basilar, and posterior cerebral arteries), respectively.

Table 3. Comparison of clinical and atherosclerotic characteristics between patients with acute ischemic stroke and nephrotic syndrome due to diabetic nephropathy (NS group) and patients with acute ischemic stroke and diabetes mellitus without nephrotic syndrome (control group)

	NS group	Control group	P
Patient (n)	7	14	
Age, y	65.3 ± 9.3	66.0 ± 8.0	—
Sex/male	6 (85.7)	12 (85.7)	—
DM	7 (100)	14 (100)	—
Duration	12.3 ± 10.1	7.9 ± 6.2	0.397
Hb A1c	8.7 ± 3.3	8.1 ± 1.7	—
on medication	4 (57.1)	7 (50.0)	—
Hypertension	5 (71.4)	10 (71.4)	1
BMI	22.9 ± 2.4	22.9 ± 3.3	0.852
AIS LAA	5 (71.4)	3 (21.4)	0.056
SVO	2 (28.6)	11 (78.6)	
AS score	5.3 ± 3.7	2.1 ± 2.0	0.019
Anterior c	2.6 ± 1.8	0.9 ± 1.3	0.035
Posterior c	2.3 ± 2.9	1.1 ± 0.9	0.588
T-CUS	2.0 ± 2.3	0.6 ± 0.9	0.133
T-EUS	4.0 ± 1.0	1.8 ± 1.1	0.031
T-ABI	1.9 ± 0.4	2.1 ± 0.3	0.459

The differences between the two groups were evaluated statistically by $\chi 2$ or Mann-Whitney's U tests.

Abbreviations: AIS – acute ischemic stroke; SVO – small-vessel occlusion; LAA – large-artery atherosclerosis; AS – atherosclerosis score; anterior c – anterior circulation; posterior c – posterior circulation; T-CUS – total carotid ultrasonic score; T-EUS – total extremities ultrasonic score; T-ABI – total ankle-brachial index.

Ultrasonography of carotid arteries and lower extremities arteries was performed and the stenotic lesions of these arteries graded. The atherosclerosis of lower extremity arteries were also evaluated by measurement of ankle-brachial index (ABI). A total carotid ultrasonic score (T-CUS) and total extremities ultrasonic score (T-EUS) were estimated as the sum of the right and left ultrasound score grades for the carotid and extremity arteries, respectively. A total ABI (T-ABI) score was estimated as the sum of the right and left ABI.

Table 4. The comparison of laboratory data between the patients of acute ischemic stroke with nephrotic syndrome due to diabetic nephropathy (NS group) and acute ischemic stroke with diabetes mellitus without nephrotic syndrome (Control group)

	NS group	Control group	P
Albumin	2.7 ± 0.5	4.2 ± 0.4	<0.001
Total cholesterol	321.7 ± 42.5	211.7 ± 39.4	0.001
LDL-cholesterol	214.4 ± 40.5	132.6 ± 34.2	0.001
HDL-choresterol	51.3 ± 17.4	49.4 ± 9.7	0.94
Triglyceride	229.0 ± 67.4	189.6 ± 203.1	0.037
Creatinin	2.1 ± 1.6	0.9 ± 0.3	0.009
eGFR	37.7 ± 23.3	71.5 ± 22.6	0.011
Hematocrit	35.7 ± 5.5	43.0 ± 5.1	0.006
Platelet	27.8 ± 2.7	22.4 ± 5.8	0.048
PT	0.9 ± 0.7	0.9 ± 0.0	0.55
APTT	28.9 ± 3.9	28.8 ± 3.1	0.933
Fibrinogen	385.4 ± 104.5	275.0 ± 57.9	0.019
D-dimer	1.8 ± 1.0	0.7 ± 0.4	0.01

The difference between two groups was statistically estimated by the method of $\chi2$ test or Mann-Whitney's U-test.

Abbreviations: LDL – low density lipoprotein; HDL – high density lipoprotein; eGFR – estimated glomerular filtration rate.

We carried out a retrospective cohort study to assess the association between NS and atherosclerosis progression. AIS patients with NS due to DMN were compared with AIS patients without NS. Cases were compared with controls in respect of body mass index (BMI), morbidity of hypertension, and several laboratory findings. Case 2 with amyloid nephropathy, case 7 with MPGN, and case 10 with CE, were excluded from this study. The remaining seven patients with AIS and NS due to DMN were selected as cases (NS group). Cases were compared with AIS patients with DM without NS (control group) who were matched to cases in a 2:1 ratio by age, sex, use of medications for DM, and HbA1c level. The control group included 11 cases with SVO and three cases with LAA, and had a larger number of SVOs than that of the NS group. The NS group had statistically significantly higher cerebral artery atherosclerosis scores, especially in the anterior circulation. The NS group also showed higher T-CUS, although there were no significant differences compared with the control group. T-EUS in the NS group were higher than those in the control group. However, there were no differences in T-ABI scores between the two groups. The NS group demonstrated hypoalbuminemia and hypercholesterolemia, as well as anemia, increased creatinine level and low eGFR, which were consistent with chronic kidney disease. They also had high fibrinogen and D-dimer levels, indicating a hypercoagulable state (Table 3, 4).

4. Cerebral Venous Sinus Thrombosis (CVT) Associated with NS

1) CVT Patients (Table 5)

Case 1: The patient was a 46-year-old man. He had headache and vomiting the day after he drank heavily. Contrast brain computed tomography (CT) and MRI revealed a defect in the transverse sinus, straight sinus, and superior sagittal sinus. His blood was hemoconcentrated, and blood test results indicated high D-dimer and fibrinogen levels and decrease of antithrombin III.

Case 2: The patient was an 89-year-old woman. She suffered from ischemic colitis. She developed left hemiplegia and headache after having had diarrhea for a month. Brain CT revealed hematoma in the subcortical region of the right frontal lobe and a high signal in the straight sinus. The superior sagittal sinus showed high-signal intensity on T1-weighted MRI and mild high-signal intensity on T2-weighted MRI. High D-dimer and fibrinogen levels were detected in the blood.

We experienced 25 cases of CVT over a period of nine years from April 2004 to August 2013. Of these cases, two had a complication of NS. The incidence of NS in CVT patients was 8.0%. Over the same period, we had 10 cases of AIS-associated NS, which is five times that of CVT associated with NS.

2) Literature Review of CVT Patients with NS

We reviewed the literatures of CVT patients with NS since 1980. The 54 cases under 20 years old were reported from Japan (20 cases) and other countries (34 cases). The 29 adult cases over 20 years old, including our two cases, were reported from Japan (10 cases) and other countries (19 cases). Table 6 shows the clinical characteristics of 18 cases (12 male and 6 female), omitting those cases for which only an abstract was available [28-42]. The average age of the cases was 44.6 ± 18.5 years (20-89 years) and minimal change disorder was detected as the cause of NS in eight cases. The clinical symptoms included headache in 14 cases (77.8%), convulsion in five cases (27.8%), motor hemiparesis in seven cases (38.9%), cranial palsies in three cases (16.7%) and aphagia in two cases (11.1%). Occluded sinuses were observed in superior sagittal sinus or transverse sinus in 13 cases (72.2%).

Table 5. The patients with cerebral venous sinus thrombosis

Case	Age	Sex	Type	MRI/CT Location of occlusion	Alb	TC	UP	day	E	DM	NS type
1	46	M	CVT	SSS, LS, straight S	2.4	364	4	3.5	+	—	minimal change
2	89	F	CVT	SSS, straight S	1.9	326	4	>3.5	+++	—	unknown

Abbreviations: TC – total cholesterol; Alb – albumin; UP – urinary protein; E – edema; DM – diabetes mellitus; NS – nephrotic syndrome; CVT – cerebral venous sinus thrombosis; SSS – superior sagittal sinus; LS – lateral sinus (transverse sinus).

NS had already been diagnosed in 13 cases before the occurrence of CVT, but in five cases was not diagnosed until the occurrence of CVT [29, 36, 39-41]. The patients showed the abnormal hypercoagulation, including decreases of antithrombin III in six cases, decrease of protein S in two cases, increased fibrinogen in 10 cases, positive lupus anticoagulant in three cases, positive antinuclear antibody in two cases, factor V Leiden mutation with hyperhomocysteinemia in one case, hemoconcentration in two cases, hormone replacement therapy with estrogen in one case, preeclampsia in one case, SLE in one case, rheumatoid arthritis (RA) in one case, ischemic colitis with diarrhea in one case, and heavy drinking in one case (Table 7).

Table 6. The adult cases of cerebral sinus thrombosis with nephrotic syndrome

Authors	year	Cases	Age/sex	Disease	Clinical pictures	Site of thrombosis	Rx	Outcome
Barthélémy	1980	1	50/M	MCD	H, papilledema	SSS, LS	AC	CR
Tovi	1988	2	46/M	n.a.	SVC syndrome	LS		CR
Levine	1989	3	32/M	MPG	H, papilledema, IV pa/sy	SSS,	AC	CR
Utunomiya	1994	4	29/M	MCNS	H, hemi-p	SSS, LS	AC	?
Burns	1995	5	20/M	MCNS	S, personality changes	SSS		Fetal
		6	35/M	MCNS	H, III palsy, hemi-p, aphagia	SS	AC	PR
Laversuch	1995	7	42/F	MGN	H, S, confusion, hemi-p	LS	AC	CR
Urch	1996	8	41/F	MCD	H, aphagia, hemi-p	SSS	AC	CR
Akatsu	1997	9	65/F	MCD	H, hemi-p	SSS		CR
Koch	1997	10	23/F	n.a.	H, visual change	SSS	AC	CR
Hirata	1999	11	46/M	MCNS	H, papilledema	SSS, LS	AC	CR
Philips	1999	12	29/M	na	S, VI palsy	SSS, LS, SS, IJV	Lysis	CR
Sung	1999	13	45/M	CTIN	H, hemi-p	SSS		CR
Nishi	2006	14	79/F	AMD	S	LS, SS		Fetal
Komada	2007	1	29/M	MCNS	H,	SSS	AC	CR
Ozkart	2011	16	56/M	MGN	H, S	LS, Straight S	AC	CR
Present cases		17	46/M	MCD	H	SSS, LS, straight S	AC	CR
		18	89/F	n.a.	H, hemi-p	SSS, straight S	AC	Fetal

Abbreviations: Rx – treatment; MCD – minimum change disease; MCNS – minimum change nephritic syndrome; MPG – mesangial proliferative glomerulonephritis; MGN – membranous glomerulonephritis; CTIN – chronic tubulointestinal nephritis; AMD – amyloidosis; n.a. – not available; H – Headache; S – seizure; hemi-p – hemiplegia; SSS – superior sagittal sinus; LS – lateral sinus (transverse sinus); SS – sigmoid sinus; IJV – internal jugular vein; AC – anticoagulation therapy; lysis – endovascular thrombolysis; CR – complete remission.

5. INTRACEREBRAL HEMORRHAGE (ICH) ASSOCIATED WITH NS

1) ICH Patients

Case 1. A 51-year-old woman was admitted to our hospital with headache and vomiting. She had right hemiparesis and speech disturbance. At age 46, she had suffered from DM, but did not take any medications. At age 47, she had double vision due to right abducent nerve palsy, and was diagnosed as having a right carotid-cavernous fistula (CCF). Endovascular coiling embolization was performed. On admission, she had slight unconsciousness (JCS I-2), dysarthria, and right side hemiparesis. NIHSS was 5 points. Her blood pressure was 170/92 mmHg. MRI showed the subcortical hemorrhage (7.1 ml) in the left temporal lobe (Figure 3 A, B). No abnormal vessel or microbleeds were detected in MRA and T2* weighted MRI. The conservative treatment was performed, but the peripheral edema and pleural effusion were not improved. On day 12, she was transferred to a specialist nephrology hospital, scoring 3 on the modified Rankin scale (mRS).

Case 2. A 63-year-old woman was admitted to our hospital because of gait disturbance. She had left hemiparesis and speech disturbance. At age 56, she had suffered from intracerebral hemorrhage (left putamen), and was under the treatment of hypertension and DM for several years. On admission, she had slight unconsciousness (JCS I-2), dysarthria, and left side hemiparesis. She also had slight residual of right hemiparesis. Bilateral pyramidal tract signs were observed. NIHSS was 6 points and her blood pressure 220/120 mmHg. A peripheral edema was observed in her lower extremities. MRI showed a hemorrhage (2.2 ml) in the right putamen, and an old lesion (slit) in the left putamen (Figure 3 C,D).

No microbleeds were detected in T2* weighted MRI. On day 16, the patient was transferred to a rehabilitation hospital on mRS 2 following treatment on the acute stage of putaminal hemorrhage.

Table 7. The hypercoagulability factors in the cases of cerebral sinus thrombosis with nephrotic syndrome

Cases	Age/sex	D-dimer[Y]	AT III	Proein S	Fbg	LA	ANA	others
1	50/M							
2	46/M				H			otitis media
3	32/M	N	N			+	—	
4	29/M	H	44.00%		H	—	—	onset on the second day of steroid pulse therapy
5	20/M		L (59 u/dl)*				—	
6	35/M		N	N	H			
7	42/F		N	17%		+	+	SLE
8	41/F							
9	65/F		33%	N		—	—	hormone (estrogen) therapy
10	23/F		N	N	H	—	—	preeclampsia, onset on the 8th day after labor
11	46/M	H	68%		H	—	—	
12	29/M					+		endovascular thrombolysis
13	45/M		N	71%	N	—	+	
14	79/F	H	N		H			RA for 20 years
15	29/M	H	L (7 mg/dl)**		H	—		
16	56/M	H	N	N	H	—	—	factor V Leiden: mutation+, Hcy-mia
17	46/M	H	80%	N	H	—	—	HC, heavy drinking
18	89/F	H			H			ischemic colitis, diarrhea

Abbreviations: Fbg – fibrinogen; LA – lupus anticoagulant; ANA – antinuclear antibody; H – high; L – low, N – normal; Hcy-mia – homocystinemia; [Y]: D-dimer or FDP (fibrin degradation products) were measured.

* normal range 80-140.

** Normal range was not described.

Table 8. Clinical characteristics in the patients of intracerebral hemorrhage with nephrotic syndrome

Case	Age	Sex	Type	MRI findings	Alb	TC	UP, g/day	E	NS type	HT	CAD	C-US rt/lt	ABI rt/lt	NIHSS at ad.	mRS at dis.	others
1	51	F	ICH	LB, temporal Lobe	2.3	363	4+	5.7	++	DMN s/p	+	—	/	/	5	3	CCF (47y)
2	63	F	ICH	Rt. putamen	2.9	348	4+	/	+	DMN s/p	+	—	/	1.15/ 1.19	6	2	ICH (61y, lt. putamen)
3	72	M	ICH	LB, multi stenosis	2.0	274	4+	/	++	AMN	—	+	0/1	1.22/ 0.85	/	5	recurrent LB
4	76	M	ICH	Cerebellum	2.7	282	4+	9.5	+	DMN s/p	+	—	¼	/	3	5	SAH (52y), NPH

Abbreviations: ICH – intracerebral hemorrhage; LB – lobar bleeding (subcortical hemorrhage); TC – total cholesterol; Alb – albumin; UP – urinary protein; E – edema; NS – nephrotic syndrome; CAD – coronary artery disease; C-US – ultrasonography of carotid artery; ABI – ankle-brachial index; NIHSS – NIH stroke scale; mRS – modified Rankin scale; CCF – carotid-cavernous fistula; DMN – diabetic nephropathy; AMN – amyloido nephropathy; s/p – suspected; ad. – admission; dis – discharge.

Table 9. Laboratory findings in the patients of intracerebral hemorrhage with nephrotic syndrome

Case	Cr	eGFR	DM, years		HbA1c	TG	HDL	LDL	Hct	Plt	PT	aPTT	Fbg	Ddimer	AT III
1	2.7	15.6	+	5	9.0	219	44	275	34.1	33.2	0.97	36.2	450	2.3	96
2	1.8	22.9	+	7	8.2	192	101	208	35.9	31.4	0.9	23.3	277	0.6	/
3	0.5	121.4	—	/	4.8	101	117	114	41.5	28.2	/	/	/	/	/
4	1.3	42.0	+	5	6.5	109	73	187	49.8	25	0.83	27.4	378	1.2	83

Abbreviations: Cr – creatinine; eGFR – estimated glomerular filtration rate; DM – diabetes mellitus; year – years of disease duration; TG: triglyceride – HDL – high density lipoprotein; LDL – low density lipoprotein; Hct – hematocrit; Plt – platelet counts; PT – prothrombin time; aPTT – activated partial thromboplastin time; Fbg – fibrinogen; AT – antithrombin.

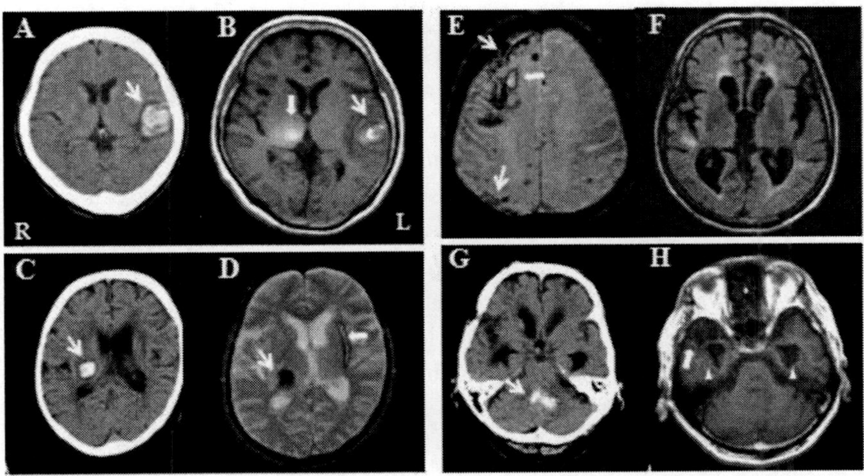

Case 1 (A, B); The arrows indicate subcortical hemorrhage (7.1 ml) in the left temporal lobe. The broad arrow is the artifact due to the coils. A, Brain CT, B; T1 weighted image, 1.5T, axial, TR; 525 ms, TE; 12 ms. Case 2 (C, D); The arrows indicate the hemorrhage (2.2 ml) in the right putamen, and the broad arrow indicates the old hemorrhagic lesion of the left putamen. C, Brain CT, D; T2* weighted image, 1.5T, axial, TR; 660 ms, TE; 25 ms. Case 3 (E, F); Broad arrow indicates subcortical hemorrhage in the right frontal lobe. Many microbleeds are shown, some of which (arrows) were clustered. Brain atrophy with ventricular dilatation and periventricular high signals are also elucidated. E; T2* weighted image, 1.5T, axial, TR; 814 ms, TE; 24.6 ms. F; T2 weighted image, 1.5T, axial, TR; 10000 ms, TE; 120 ms. Case 4 (G, H); The hemorrhage (arrow) in the deep part of cerebellar hemisphere, the old infarct (broad arrow) in the polar head of right temporal lobe and the ventricular dilatation (arrow heads) are shown. G; Brain CT, H; T1 weighted image, 1.5T, axial, TR; 530 ms, TE; 12 ms.

Figure 3. CT and MRI of the patients with intracerebral hemorrhage.

Case 3. A 72-year-old man was admitted to our hospital because of generalized convulsion. He had recurrent episodes of lobar hemorrhage (subcortical hemorrhage). He was bed-ridden after 10 episodes over a period of several years. He had vascular dementia (MMSE; 7/30 points). On admission, he had quadriplegia and bulbar palsy, showing bilateral pyramidal signs. He showed dyspnea due to pneumonia and marked peripheral edema in his extremities. His blood pressure was 104/70 mmHg, He had no DM. MRI showed a subcortical hemorrhage in right frontal lobe, and many microbleeds in bilateral frontal, temporal, parietal and occipital lobes, some of which presented as accumulations or clusters (Figure 3 E,F). MRA showed multiple stenosis of intracerebral arteries. The patient was diagnosed with cerebral amyloid angiopathy. He was treated by methyl prednisolone pledonisorone, but the excretion of urinary protein was not improved. Two months later, he was transferred to a specialist specialized nephrology hospital on mRS 5.

Case 4. A 76-year-old man was admitted to our hospital because of vertigo. At age 52, he had a subarachnoidal hemorrhage and underwent an operation for a ruptured aneurysm. For several years, he received treatment for hypertension and DM, which were not well controlled. On admission, he had cerebellar ataxia and dysarthria, but no meningeal irritation or pyramidal signs. NIHSS was 3 points. His blood pressure was 200/80 mmHg. He showed peripheral edema in his lower extremities. MRI showed the hemorrhage (3.1 ml) in the deep part of left cerebellar hemisphere with little extravasation in ventricles. An old infarct lesion in the right temporal lobe and ventricular dilatation were recognized (Figure 3 G, H). On day 26, he was transferred to a rehabilitation hospital on mRS 5 after treatment on the acute stage of the cerebellar hemorrhage.

Tables 8 and 9 show the clinical characteristics of, and laboratory findings for the patients with ICH with NS. Increases of HbA1c, fibrinogen, d-dimer and creatinine were found in three cases, two cases, two cases and three cases, respectively. Serum level of antithrombin III was normal in two cases examined. ABI was abnormal in one case, and ultrasonography of carotid artery detected mild to moderate stenosis of internal carotid artery in two cases.

The basic diseases underlying NS were suggested as amyloid nephropathy in Case 3 and DMN in Cases 1, 2 and 4. The incidence of ICH associated with NS was 0.18% of total stroke or 0.036% of ICH.

DISCUSSION

Patients with NS have a high incidence (21 to 51% of patients) of venous thrombosis, particularly deep vein and renal vein thrombosis, and pulmonary emboli, especially in younger patients under 20 [43-48], and CVT has also (rarely) been reported [2]. On the other hand, the relative risk of arterial thrombosis is low (1.0 to 5.5%), compared to that of venous thrombosis [3, 49, 50]. In 2014, Sasaki et al. reported an additional case of AIS with NS, and reviewed 21 prior cases reported in 19 literatures [4]. However, the incidence and clinical characteristics of AIS patients have remained unclear [5-10]. We reported six LAA patients, three SVO patients and one CE patient, and elucidated the following. The incidence of AIS associated with NS was 0.09%

of total stroke or 0.12% of AIS, and five times that of CVT in the adult patients with NS. The patients had severe stenosis and/or occlusion of intracranial cerebral arteries, especially in the cerebral anterior circulation, and also demonstrated atherosclerosis of the internal carotid and lower extremity arteries. The patients had high serum fibrinogen and D-dimer levels, suggesting that AIS patients with NS have a greater degree of hypercoagulability than AIS patients without NS. The pathogenesis of NS was from DMN in eight cases (80%), amyloid nephropathy in one case, and MPGN in one case. These results suggest that the AIS patients with NS, especially from DMN, had marked general arteriosclerosis of the body, which might be a clinical characteristic peculiar to Japanese patients [24].

Mahmoodi et al. reported that annual incidences of venous and arterial thomboembolism in NS patients were 1.02% and 1.48%, respectively – eight times higher than in general population from a retrospective cohort study. They also reported that multiple classic risk factors for atherosclerosis are associated with arterial thromboembolism in NS patients, including sex, age, hypertension, DM, smoking, prior arterial thromboembolism, and eGFR. The annual incidence of DMN patients was 7.43, which was the highest incidence among all types of nephropathy [3].

Sasaki et al. reported that histopathology of NS in the 22 AIS patients previously reported from several countries was membranous nephropathy in five cases (22.7%), minimal change in four cases (18.2%), MPGN in three cases (13.6%), focal segmental sclerosis in two cases (9.0%), Ig A nephropathy, undetermined pathology in six cases (27.2%), and DMN in only one case (4.5%) [4]. In our study, DMN was clinically apparent for the pathogenesis of NS in 80% of patients. The occurrence of non-diabetic renal disease (NDRD) in DMN patients has been increasingly recognized in recent years. Zhuo et al. reviewed the 13 literatures from the period between 1983 and 2012, and reported the prevalence of DMN complicating NDRD; the common histological diagnoses were tubule-interstitial nephritis (22.7%) and IgA nephropathy (14.1%) [51]. The complication of NDRD notwithstanding, DMN could be the primary cause of the pathogenesis of NS in 80% of our patients. The patients with NS had the combined risks from the basic, disease-induced NS, as well as the hypercoagulation state arising as a result of NS. It is well known that DM and hyperlipidemia are strong risk factors for AIS, but it is possible that the existence of NS may be another risk factor. DM may be the strongest and most important risk factor for AIS associated with NS, especially in Japanese patients.

CVT commonly occurs in young adults under 40 years old, and is around three times more common in females than in males. Its incidence is about 0.5-1.0% among all types of stroke. There are many risk factors of CVT, one of which is NS [2, 11, 12, 52-54]. Twenty-five patients (aged 59.8 ± 18.6 years; 20 men and five women) were admitted to our center during a nine-year period between April 2004 and August 2013 [55]. The incidence of CVT in our study was 0.22% of all kinds of strokes or 0.30% of ischemic infarctions, which may be lower than previously reported. The patients with CVT in our study had, as risk factors, oral contraceptive use (two patients), deficiency of protein S (three patients) and antithrombin (two patients), nephrotic syndrome (two patients), iron deficiency anemia (three patients), colon cancer (one patient), and multiple myeloma in remission (one patient). Furthermore, hyperhomocysteinemia was identified in nine (56.3%) of 16 patients tested (Table 10). Our results show that the CVT patients in Japan may be older and include fewer females than CVT patients in Western countries [55]. The most common risk factor in women has been reported to be oral contraceptive use [2, 11, 12, 14, 15, 16]. Oral contraceptive use has been reported to be less prevalent as a cause of CVT in Japan than in Western countries [56]. This may be due to few young female CVT patients using oral contraceptives in Japan.

NS is one of the risk factors for CVT [28, 53, 54]. A multinational (21 countries), multicenter (89 centers) prospective observational study entitled 'The International Study on Cerebral Vein and Dural Sinus Thrombosis' (ISCVT) [52, 57-60] examined 624 cases of CVT. ISCVT and other studies reported that the prevalence of NS was 0.6-2% [12, 52, 54, 61]. In our study, NS was recognized as a risk factor for CVT in two out of 25 cases (8.0%) [55]. Our study was not performed in multiple centers, but in only one center. In the future, a nationwide prospective study needs to be performed in order to clarify the characteristics of CVT in more detail, especially its risk factors, in Japanese patients of all ages. Patients with NS have a thrombotic tendency; both venous thrombosis and arterial thrombosis may occur. In the literature, the number of published cases of CVT was tenfold that of cerebral artery thrombosis as a complication of NS in individuals aged <20 years. In adults, however, the number of cases of CVT reported was twice that of cerebral artery thrombosis cases. Venous thrombosis is likely to occur in young cases with NS, and both venous and arterial thromboses are likely to occur in adult cases [25]. NS shows a thrombotic tendency, but the onset of CVT may develop as a result of some additional thrombotic factor, such as hemo-concentration after heavy drinking, diarrhea, or vomiting. Lifestyle management alongside the conventional treatment of NS is important.

It is well known that hemorrhagic lesions possibly occur as an associated complication of CVT [2]. However, the clinical details of ICH-associated NS have not been reported [62]. Patients with NS have a thrombotic tendency, which, as described above, is a risk factor for both venous thrombosis and arterial thrombosis. When we retrospectively analyzed the hospitalized patients, however, we not only found the 10 AIS and 2 CVT patients with NS, but also, unexpectedly, four ICH patients with NS [26]. The incidence of ICH associated with NS was 0.18% of total stroke or 0.036% of ICH. The basic disorders associated with NS were DMN and amyloidosis in three cases and one case, respectively. These patients had strong risk factors for ICH, such as hypertension, vascular vulnerability with atherosclerosis or amyloidosis, with change of bloodstream after operation, suggesting that they had overcome the risk of thrombophilia (Figure 4). NS essentially associates with a risk for venous or arterial thrombosis. Attention should also be paid to NS as a risk factor for ICH.

The main cause of NS in children is minimal change disease (MCD). On the other hand, approximately 30% of adults with NS have a systemic disease, such as DM, amyloidosis, or SLE. The remaining cases are usually due to primary disorders including MCD, focal segmental glomerulosclerosis (FSGS), and membranous nephropathy. The frequency of the different forms of nephropathy underlying NS in adults was evaluated in renal biopsy registries of several counties [63-70]. Among patients between the ages of 15 and 65, the most common causes of NS were membranous nephropathy, MCD, lupus nephritis, FSGS, MPGN, amyloidosis, and IgA nephropathy. A similar distribution was observed among elderly individuals (aged above 65 years), except for an increased prevalence of amyloidosis and a decreased prevalence of lupus. The frequency of the DMN in these registries was ~1-5% [63-70].

In Japan, the frequency of nephropathy underlying NS was pathologically evaluated in the Japan Renal Biopsy Registry (J-RBR) of 2,751 patients. Among patients between 20 and 65 years of age, the most common causes of NS were membranous nephropathy (18%), minimal change disease (25%), DMN (12%), lupus nephritis (10%), FSGS (7%), MPGN (2%), amyloidosis (2%), and IgA nephropathy (7%). Among the elderly individuals (age greater than 65 years), the most common causes of NS were membranous nephropathy (32%), MCD (13%), DMN (10%), lupus nephritis (2%), FSGS (6%), MPGN (4%), amyloidosis (8%), and IgA nephropathy (4%). The most remarkable characteristic form underlying NS in the J-RBR evaluation was the higher frequency of DMN than those in the other registries [71, 72]. Furthermore, it has recently been reported that 41.4% of 614 patients with pathologically confirmed DMN registered in J-RBR were in a clinically nephrotic state [73]. In our stroke registry, DMN was the main disorder underlying NS in both AIS and ICH associated with NS (Figure 5). Long-term DM could damage the arterial vessels so as to precipitate the occurrence of both diseases. On the other hand, DM cannot significantly damage the veins, indicating that the thrombotic tendency due to NS may be the main player in causing CVT. NS tends to be associated with a risk factor for AIS and CVT, and attention must also be paid to ICH associated with NS in cases of stroke.

Table 10. The risk factors and etiology in the patients with cerebral venous sinus thrombosis

Patients	n = 25	%
Sex (male)	20	80
Age		
Male	59.8 ± 17.6	
Female	59.6 ± 24.4	
Lab. findings and risk factors		
d-dimer n = 20	20	100
Protein C deficiency n = 15	0	0
Protein S deficiency n = 15	3	20
Antithrombin deficiency n = 14	2	14.3
ANA n = 14	0	0
Anti-phospholipid syndrome n = 15	0	0
CRP n = 21	19	90.5
Fibrinogen n = 18	14	77.8
Anemia (fe def.) n = 25	3	12
Homocysteine n = 16	9	56.3
Oral contraceptives (female n = 5)	2	8 (40)
Nephrotic syndrome	2	8
Malignancy	1	4
Others (associated)		
Multiple myeloma	1	
Multiple sclerosis	1	
Neurofibromatosis 1	1	
Epilepsy	1	

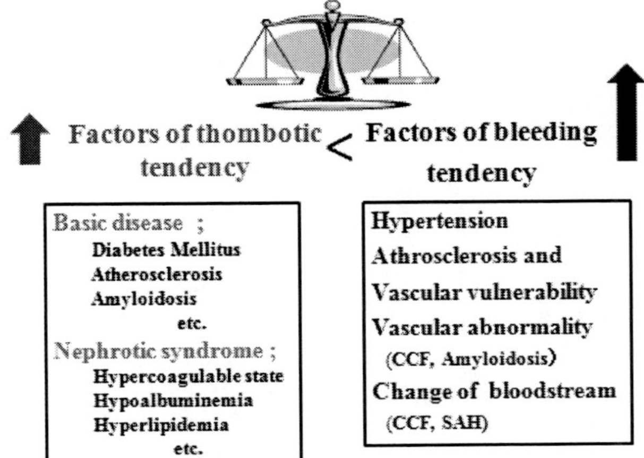

CCF: carotid-cavernous fistula.

Figure 4. The factors of thrombotic tendency and the factors of bleeding tendency in the patients with intracerebral hemorrhage associated with nephrotic syndrome.

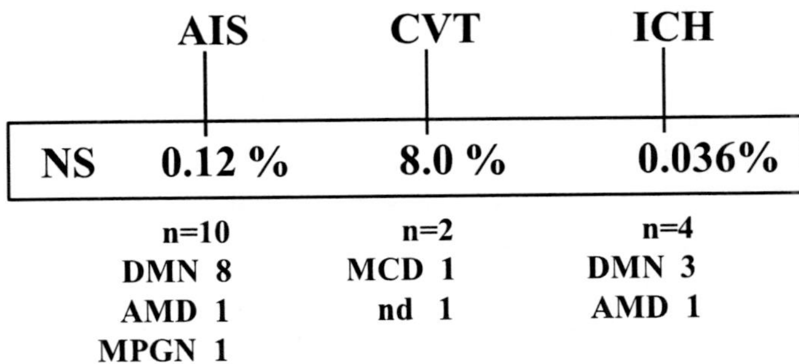

Figure 5. Nephrotic syndrome (NS) and diabetic nephropathy (DMN) in the patients with acute ischemic stroke (AIS), cerebral venous sinus thrombosis (CVT) and intracerebral hemorrhage (ICH).
AMD – amyloido nephropathy; MPGN – membranoproliferative glomerulonephritis; MCD – minimal change disease.

REFERENCES

[1]   Crew RJ, Radhakrishnan J, Appel G. Complications of the nephrotic syndrome and their treatment. Clin Nephrol 2004;62:245.

[2]   Bousser M-G, Ferro JM. Cerebral venous thrombosis: an update. Lancet Neurol. 2007;6:162-170.

[3]   Mahmoodi BK, ten Kate MK, Waanders F et al. High absolute risks and predictors of venous and arterial thromboembolic events in patients with nephrotic syndrome: results from a large retrospective cohort study. Circulation. 2008;117:224-230.

[4]   Sasaki Y, Raita Y, Uehara G et al. Carotid thromboembolism associated with nephrotic syndrome treated with dabigatran. Case Rep. Nephrol. Urol. 2014;4:42-52.

[5]   Leno C, Pascual J, Polo JM et al. Nephrotic syndrome, accelerated atherosclerosis, and stroke. Stroke. 1992;23:921-922.

[6]   de Gauna RR, Alcelay LG, Conesa MJ et al. Thrombosis of the posterior inferior cerebellar artery secondary to nephrotic syndrome. Nephron. 1996;72:123.

[7]   Nandish SS, Khardori R, Elamin EM. Transient ischemic attack and nephrotic syndrome: Case report and review of literature. Am J Med Sci. 2006;332:32-35.

[8]   Yeh SM, Lee JJ, Hung CC et al. Acute cerebral infarction in a patient with nodular glomerulopathy--atypical features and differential diagnosis. Kaohsiung J Med Sci. 2011;27:39-44.

[9]   Babu A, Boddana P, Robson S et al. Cerebral infarction in patient with minimal change nephrotic syndrome. Indian J Nephrol. 2013;23:51-53.

[10]   Gigante A, Barbano B, Liberatori M et al. Nephrotic syndrome and stroke. Int J Immunopathol Pharmacol. 2013;26:769-772.

[11]   Stam J. Thrombosis of the cerebral veins and sinuses. N Engl J Med. 2005;28;352:1791-1798.

[12]   Saposnik G, Barinagarrementeria F, Brown RD Jr et al. Diagnosis and management of cerebral venous thrombosis: a statement for healthcare professionals from the American Heart Association/American Stroke Association. Stroke. 2011;42:1158-1192.

[13]   Makris M. Hyperhomocysteinemia and thrombosis. Clin Lab Haematol. 2000;22:133-143.

[14]   de Bruijn SF, Stam J, Koopman MM, Vandenbroucke JP. Case-control study of risk of cerebral sinus thrombosis in oral contraceptive users and in [correction of who are] carriers of hereditary prothrombotic conditions. The Cerebral Venous Sinus Thrombosis Study Group. BMJ 1998;316:589.

[15]   Saadatnia M, Fatehi F, Basiri K et al. Cerebral venous sinus thrombosis risk factors. Int J Stroke. 2009;4:111-123.

[16] Martinelli I, Sacchi E, Landi G et al. High risk of cerebral-vein thrombosis in carriers of a prothrombin-gene mutation and in users of oral contraceptives. N Engl J Med 1998;338:1793-1797.

[17] Kauffmann RH, Veltkamp JJ, Van Tilburg NH et al. Acquired antithrombin III deficiency and thrombosis in the nephrotic syndrome. Am J Med 1978;65:607-613.

[18] Vaziri ND, Paule P, Toohey J et al. Acquired deficiency and urinary excretion of antithrombin III in nephrotic syndrome. Arch Intern Med 1984;144:1802-1803.

[19] Singhal R, Brimble KS. Thromboembolic complications in the nephrotic syndrome: pathophysiology and clinical management. Thromb Res 2006;118:397-407.

[20] Joven J, Villabona C, Vilella E. et al. Abnormalities of lipoprotein metabolism in patients with the nephrotic syndrome. N Engl J Med 1990;323:579-584.

[21] Stenvinkel P, Berglund L, Heimbürger O et al. Lipoprotein(a) in nephrotic syndrome. Kidney Int 1993;44:1116-1123.

[22] Gigante A, Barbano B, Sardo L et al. Hypercoagulability and nephrotic syndrome. Curr Vasc Pharmacol. 2014;12:512-517.

[23] Barbano B, Gigante A, Amoroso A et al. Thrombosis in nephrotic syndrome. Semin Thromb Hemost. 2013;39:469-476.

[24] Iwaki H, Kuriyama M, Neshige S et al. Acute ischemic stroke associated with nephrotic syndrome: Incidence and significance-Retrospective cohort study. eNeurologicalSci. e Neurological. Sci. 2015;1:47-50.

[25] Iwaki H, Neshige S, Hara N et al. Cerebral venous thrombosis as a complication of nephrotic syndrome — A case report and literature review. Clin Neurol (Japanese) 2014;54:495-501.

[26] Kono R, Iwaki H, Takeshima S et al. Intracerebral hemorrhage associated with nephrotic syndrome — Prevalence and clinical characteristics. Clin Neurol (Japanese) 2016;56:180-185.

[27] Uekita K, Hasebe N, Funayama N et al. Cervical and intracranial atherosclerosis and silent brain infarction in Japanese patients with coronary artery disease. Cerebrovasc Dis. 2003;16:61-68.

[28] Barthélémy M, Bousser MG, Jacobs C. Cerebral venous thrombosis, complication of the nephrotic syndrome. Nouv Presse Med 1980;9:367-369.

[29] Tovi F, Hirsch M, Gatot A. Superior vena cava syndrome: presenting symptom of silent otitis media. J Laryngol Otol 1988;102:623-625.

[30] Levine SR, Kieran S, Puzio K et al. Cerebral venous thrombosis with lupus anticoagulants. Report of two cases. Stroke 1987;18:801-804.

[31] Utunomiya Y, Uzawa N, Inafuku T et al. A case of minimal change nephrotic syndrome complicated by cerebral sinus thrombosis. Jin to Touseki (Japanese) 1994;37:805-809.

[32] Burns A, Wilson E, Harber M et al. Cerebral venous sinus thrombosis in minimal change nephrotic syndrome. Nephrol Dial Transplant 1995;10: 30-34.

[33] Laversuch CJ, Brown MM, Clifton A et al. Cerebral venous thrombosis and acquired protein S deficiency: an uncommon cause of headache in systemic lupus erythematosus. Br J Rheumatol 1995;34:572-575.

[34] Urch C, Pusey CD. Sagittal sinus thrombosis in adult minimal change nephrotic syndrome. Clin Nephrol 1996;45:131-132.

[35] Akatsu H, Vaysburd M, Fervenza F et al. Cerebral venous thrombosis in nephrotic syndrome. Clin Nephrol 1997;48:317-320.

[36] Koch CA, Robyn JA, Walz ET et al. Cranial sinus thrombosis and preeclampsia. J Stroke Cerebrovasc Dis 1997;6:430-433.

[37] Hiraa M, Kuroda M, KonI. Cerebral venous thrombosis in minimal change nephrotic syndrome. Nihon Jinzo Gakkai Shi (Japanese) 1999; 41:464-468.

[38] Philips MF, Bagley LJ, Sinson GP et al. Endovascular thrombolysis for symptomatic cerebral venous thrombosis. J Neurosurg 1999;90:65-71.

[39] Sung SF, Jeng JS, Yip PK et al. Cerebral venous thrombosis in patients with nephrotic syndrome--case reports. Angiology 1999;50:427-432.

[40] Nishi H, Abe A, Kita A et al. Cerebral venous thrombosis in adult nephrotic syndrome due to systemic amyloidosis. Clin Nephrol 2006;65: 61-64.

[41] Komaba H, Kadoguchi H, Igaki N et al. Early detection and successful treatment of cerebral venous thrombosis associated with minimal change nephrotic syndrome. Clin Nephrol 2007;68:179-181.

[42] Ozkurt S, Temiz G, Saylisoy S et al. Cerebral sinovenous thrombosis associated with factor V Leiden and methylenetetrahydrofolate reductase A1298C mutation in adult membranous glomerulonephritis. Ren Fail 2011;33:524-527.

[43] Llach F. Hypercoagulability, renal vein thrombosis, and other thrombotic complications of nephrotic syndrome. Kidney international. 1985;28: 429-439.

[44] Llach F, Papper S, Massry SG. The clinical spectrum of renal vein thrombosis: acute and chronic. Am J Med 1980;69:819-27.

[45] Chugh KS, Malik N, Uberoi HS, Gupta VK, Aggarwal ML, Singhal PC et al. Renal vein thrombosis in nephrotic syndrome—a prospective study and review. Postgrad Med J 1981;57:566-70.

[46] Velasquez FF, Garcia PN, Ruiz MN. Idiopathic nephrotic syndrome of the adult with asymptomatic thrombosis of the renal vein. Am J Nephrol 1988;8:457-62.

[47] Wagoner RD, Stanson AW, Holley KE, Winter CS. Renal vein thrombosis in idiopathic membranous glomerulopathy and nephrotic syndrome: incidence and significance. Kidney Int 1983;23:368-74. Int. 1993;44:1116-1123.

[48] Bennett WM. Renal vein thrombosis in nephrotic syndrome. Ann Intern Med 1975;83:577-8.

[49] Ordoñez JD, Hiatt RA, Killebrew EJ et al. The increased risk of coronary heart disease associated with nephrotic syndrome. Kidney Int. 1993;44:638-642.

[50] Wass VJ, Jarrett RJ, Chilvers C et al. Does the nephrotic syndrome increase the risk of cardiovascular disease? Lancet. 1979;2(8144):664-667.

[51] Zhuo L, Zou G, Li W et al. Prevalence of diabetic nephropathy complicating non-diabetic renal disease among Chinese patients with type 2 diabetes mellitus. Eur J Med Res. 2013; 18:4.

[52] Ferro JM, Canhão P, Stam J et al. Prognosis of cerebral vein and dural sinus thrombosis: results of the International Study on Cerebral Vein and Dural Sinus Thrombosis (ISCVT). Stroke 2004;35:664-670.

[53] Ferro JM, Canhão P, Stam J et al. Delay in the diagnosis of cerebral vein and dural sinus thrombosis: influence on outcome. Stroke 2009;40:3133-3138.

[54] Wasay M, Bakshi R, Bobustuc G et al. Cerebral venous thrombosis: analysis of a multicenter cohort from the United States. J Stroke Cerebrovasc Dis 2008;17:49-54.

[55] Takemaru M, Kuriyama M, Himeno T et al. Cerebral venous sinus thrombosis: Incidence and hyperhomocysteinemia as a risk factor in Japanese patients. In preparation On submission.

[56] Sato R, Iwasawa M. Contraceptive use and induced abortion in Japan: how is it so unique among the developed countries? Jpn J Popul. 2006; 4:33-54.

[57] Ferro JM, Canhão P, Bousser MG et al. Cerebral vein and dural sinus thrombosis in elderly patients. Stroke 2005;36:1927-1932.

[58] Girot M, Ferro JM, Canhão P et al. Predictors of outcome in patients with cerebral venous thrombosis and intracerebral hemorrhage. Stroke 2007;38:337-342.

[59] Coutinho JM, Ferro JM, Canhão P et al. Cerebral venous and sinus thrombosis in women. Stroke 2009;40:2356-2361.

[60] Coutinho JM, Stam J, Canhão P et al. Cerebral venous thrombosis in the absence of headache. Stroke 2015;46:245-247.

[61] Lauw MN, Barco S, Coutinho JM, Middeldorp S. Cerebral venous thrombosis and thrombophilia: a systematic review and meta-analysis. Semin Thromb Hemost 2013;39:913.

[62] Hu P, Zhao XQ, Hu B et al. Spontaneous intracerebral hemorrhage in a pediatric patient with nephrotic syndrome. J Clin Hypertens (Greenwich). 2014;16:236-237.

[63] Rivera F, López-Gómez JM, Pérez-García R et al. Clinicopathologic correlations of renal pathology in Spain. Kidney Int. 2004;66:898-904.

[64] Haas M, Meehan SM, Karrison TG, Spargo BH. Changing etiologies of unexplained adult nephrotic syndrome: a comparison of renal biopsy findings from 1976-1979 and 1995-1997. Am J Kidney Dis 1997;30:621.

[65] Braden GL, Mulhern JG, O'Shea MH et al. Changing incidence of glomerular diseases in adults. Am J Kidney Dis 2000; 35:878.

[66] Simon P, Ramee MP, Boulahrouz R et al. Epidemiologic data of primary glomerular diseases in western France. Kidney Int 2004; 66:905.

[67] Malafronte P, Mastroianni-Kirsztajn G, Betônico GN et al. Paulista Registry of glomerulonephritis: 5-year data report. Nephrol Dial Transplant 2006; 21:3098.

[68] Bahiense-Oliveira M, Saldanha LB, Mota EL et al. Primary glomerular diseases in Brazil (1979-1999): is the frequency of focal and segmental glomerulosclerosis increasing? Clin Nephrol 2004;61:90.

[69] Gesualdo L, Di Palma AM, Morrone LF et al. The Italian experience of the national registry of renal biopsies. Kidney Int 2004;66:890.

[70] Heaf J. The Danish Renal Biopsy Register. Kidney Int 2004; 66:895.

[71] Yokoyama H, Sugiyama H, Sato H et al. Renal disease in the elderly and the very elderly Japanese: analysis of the Japan Renal Biopsy Registry (J-RBR). Clin Exp Nephrol. 2012;16:903-920.

[72] Sugiyama H, Yokoyama H, Sato H et al. Japan Renal Biopsy Registry and Japan Kidney Disease Registry: Committee Report for 2009 and 2010. Clin Exp Nephrol. 2013;17:155-173.

[73] Wada T. Progressive renal diseases: Recent advances in diagnosis and treatments. Diabetic nephropathy. J Jap Soc Int Med 2013 (Japanese); 102:1152-1158.

INDEX

A

acetylcholine, 14, 75, 82
acromegaly, 15
adenoma, 16, 146
adhesions, 58
adults, 13, 124, 154, 158, 159, 213, 214, 215, 222
aggregation, 199
agonist, 80, 145
akinesia, 157
albumin, 200, 201, 202, 206, 210
ALS, 91, 106
American Heart Association, 218
amplitude, 35, 37, 38, 39, 40, 41, 42, 43
amygdala, 6, 7, 21
amyloidosis, xiii, 198, 202, 207, 214, 215, 220
anatomy, 20, 23
anemia, 205, 214
aneurysm, 50, 51, 52, 57, 212
angiography, 18, 50, 52, 57, 185
angioma, 144
ankylosis, 161
ANOVA, 96, 101
anterior cingulate cortex, 21, 153
antibody, x, 86, 90, 92, 93, 99, 101, 112, 117, 122, 156, 206, 209
anticholinergic, 151
anticoagulant, 200, 206, 209

anticoagulation, 207
antidiuretic hormone, 151
antipsychotic, 15
antisense, 71
antiviral drugs, 15
anxiety, 19, 161
aphasia, 111, 128, 175
apnea, 157
apoptosis, x, 60, 62, 63, 69, 71, 72, 73, 74, 76, 77, 83, 84, 85
appetite, 9
arginine, 69, 71, 150
arterial vessels, 215
arteries, xii, 183, 197, 198, 200, 202, 203, 204, 211, 213
arteriosclerosis, 213
artery, ix, xii, 49, 51, 55, 138, 167, 170, 173, 174, 175, 177, 182, 184, 188, 191, 192, 198, 201, 202, 203, 205, 210, 212, 214, 217, 219
arthritis, 207
aseptic meningitis, xi, 123, 125, 127, 128, 129, 133
astrocytes, 73, 112, 114
asymptomatic, 152, 220
ataxia, 139, 148, 149, 159, 212
atherosclerosis, xii, 167, 182, 198, 200, 201, 202, 203, 204, 213, 214, 217, 219
ATP, 69, 83
atrial fibrillation, 184

atrophy, 91, 102, 129, 135, 136, 138, 141, 143, 145, 146, 147, 148, 149, 150, 151, 152, 154, 159, 160, 163, 164, 165, 211
autonomic neuropathy, 144
autopsy, 50, 115, 117, 120
autosomal dominant, 92
axons, x, xii, 6, 11, 13, 14, 60, 67, 68, 75, 76, 80, 114, 117, 120, 167, 170

B

basal forebrain, 15
basal ganglia, 6, 202
basilar artery, 203
Bcl-2 family, 71
BDNF, ix, 60, 64, 65, 66, 68, 70, 73, 77, 78, 79, 80
behaviors, 3, 14
benign, 16, 129, 130, 131, 132, 133, 145, 149, 151, 159
benign prostatic hyperplasia, 145
benign tumors, 16
bilateral, 117, 121, 148, 211
biocompatibility, ix, 50, 55, 56, 58
biopsy, 110, 111, 112, 113, 114, 115, 116, 117, 119, 120, 202, 215, 222
bleeding, 129, 185, 210, 216
blood pressure, ix, 49, 55, 56, 208, 211, 212
blood vessels, 11, 69
blood-brain barrier (BBB), 6, 114, 116
bloodstream, xiii, 198, 215
body mass index (BMI), 203, 204
bowel, 126, 150, 151, 154, 156, 159, 163, 164, 165
brachial plexus, ix, 59, 60, 78, 81, 82
brain abscess, 169
brain activity, 11
brain damage, vii, 1, 4, 169
brain functions(ing), viii, 2, 5, 17
brain herniation, 14, 199
brain stem, 6, 12, 13, 14, 15
brain structure, 22
brain tumor, 12, 18, 169
brainstem, xii, 129, 140, 157, 167, 170, 180, 184, 187, 194

breakdown, ix, 50, 55, 56, 199

C

cadaver, 18
cancer, 3, 17, 214
capillary, 199
capsule, xii, 19, 138, 167, 170, 180, 184, 193, 194
cardiac arrhythmia, 182
cardiovascular disease, 221
carotid arteries, 203, 204
carotid endarterectomy, 186
case studies, 179, 185
caspases, 62, 71, 74, 84, 85
cauda equina, 77, 87, 125, 129, 134
causal relationship, 42
CD8+, 110, 112, 113, 114, 116, 118, 119
cell biology, 5, 86
cell body, ix, 10, 59
cell death, x, 60, 71, 72, 76, 79, 83, 84, 100
cell fate, 64, 72
cell line, ix, 60, 64, 65, 76, 78, 79, 80, 87
cell membranes, 62
cell surface, 71
central nervous system (CNS), viii, ix, x, 2, 15, 16, 18, 59, 67, 68, 72, 76, 85, 109, 110, 114, 117, 118, 119, 121, 124, 125, 128, 129, 169
central pontine myelinolysis, 147
cerebellum, 6
cerebral amyloid angiopathy, 212
cerebral aneurysm, 51, 52, 56, 57
cerebral arteries, 202, 203, 213
cerebral cortex, 7, 12, 19, 23, 92, 95, 99, 190
cerebral hemisphere, 6, 11, 169
cerebrospinal fluid (CSF), xi, 111, 117, 124, 125, 126, 128, 129, 133,159
cerebrovascular disease, 184
cerebrovascular neurosurgery, vii, ix, 50, 56
ChAT, 75
chemical reactivity, 69
children, 121, 124, 135, 142, 159, 215
cholesterol, 200, 201, 202, 204, 206, 210

choline, 75, 86

chondroitin sulfate, 85

chromosome, 90, 92, 103, 104, 105, 106

chronic kidney disease, 205

circadian rhythm, 146, 150

circulation, xiii, 7, 198, 203, 205, 213

clinical examination, 169

clinical presentation, 110, 115

clinical symptoms, vii, xi, 124, 125, 206

clustering, 102, 112, 116

clusters, x, 90, 92, 96, 97, 98, 99, 100, 101, 111, 211

cobalt, 51

cocaine, 15

coding, 29

cognition, 7, 8, 15

cognitive abilities, 8

cognitive domains, 37

cognitive function, 11, 160, 163

cognitive impairment, 160

cognitive process, 42

cognitive psychology, 11

colitis, 205, 207, 209

collective unconscious, 12

colon cancer, 214

coma, 14

combination therapy, 162

compaction, 62, 170

complications, xii, 165, 168, 197, 198, 218, 220

computed tomography, xii, 168, 173, 205

COMT inhibitor, 162

conduction, 117, 118, 122, 125, 129

connective tissue, 60

consciousness, 126, 128, 158

constipation, 145, 148, 149, 153, 156, 157

contraceptives, 199, 214, 216, 218

control group, 66, 67, 70, 203, 205

contusion, 77

convergence, 5, 79

convulsion, 206, 211

coronary artery disease (CAD), 210, 219

coronary heart disease, 220

corpus callosum, 6, 93

correlation(s), 29, 31, 32, 33, 41, 82, 175, 178, 182, 188, 191, 193, 194, 221

cortex, x, 7, 8, 11, 12, 14, 19, 20, 21, 23, 89, 91, 92, 95, 98, 99, 112, 116, 147, 153, 169, 174, 175, 178, 183, 186, 189, 190, 191

corticobasal degeneration (CBD), 96, 102, 107, 144

creatinine, 205, 210, 212

criticism, 17

CRP, 216

CSPG, 72, 73

CST, 179, 180, 181

CT and MRI artifacts, ix, 49

CT scan, 185

CTA, 50

cytoarchitecture, 10

cytochrome, 69, 71

cytokines, viii, 2, 15

cytomegalovirus, 156

cytoplasm, 62, 74, 90, 100, 113

D

deep brain stimulation, 15, 18, 19, 25

defecation, 147

defects, 92, 93

deficiency, 150, 214, 216, 218, 219

deficit, 21, 169

deformation, 55

degenerate, 15

degradation, 209

dehydration, 158, 199

dementia, 91, 93, 103, 104, 105, 106, 107, 144, 146, 149, 152, 158, 159, 161, 211

demyelinating disease, x, 109, 119, 120

demyelination, xi, 109, 110, 111, 112, 113, 114, 115, 116, 117, 118, 119, 120, 121, 125, 128

dendrites, 7, 11, 75

deposition, 111, 114

depression, 12, 159

deprivation, 83

derivatives, 69

developing brain, 8

diabetes, 147, 150, 184, 203, 204, 206, 210, 221

diabetic nephropathy, xiii, 198, 202, 203, 204, 210, 217, 221

diabetic neuropathy, 15, 129

diagnostic criteria, 93

diarrhea, 205, 207, 209, 214

differential diagnosis, xi, 124, 129, 168, 184, 218

diffusion, 168, 169, 173, 174, 179, 189, 190, 192, 193, 194

diffusion-weighted imaging (DWI), 169, 170, 174, 178, 181, 184, 187, 192

DISC, 71

diseases, xii, 66, 69, 72, 104, 119, 124, 132, 133, 147, 149, 151, 157, 164, 197, 198, 199, 202, 212, 215, 222

disorder, xi, xiii, 90, 92, 110, 124, 127, 131, 157, 198, 206, 215

dizziness, 145

DNA, x, 63, 74, 77, 89, 90, 104, 105, 106, 133

DNA damage, 63

DOC, 126

DOI, 22, 58, 120, 161

dopamine, 14, 20, 142, 146

dopaminergic, 14, 65, 78, 142

dosage, 66, 71, 74

double-blind trial, 140

down-regulation, 75

drug delivery, 66

drugs, 15, 74

dysarthria, 169, 181, 193, 208, 212

dyspnea, 211

E

E. coli, 141

edema, xii, 197, 198, 200, 202, 206, 208, 210, 211, 212

EEG, viii, 27, 29, 30, 31, 33, 34, 35, 41, 42, 46, 47, 48

EEG activity, 30

elderly population, 151

electric current, 3

electrodes, 30, 31, 34, 39, 40, 41, 43

electromyography (EMG), 68, 127, 145, 146, 152, 158, 165

electron, 52, 58, 117

electron microscopy, 58, 117

emboli, 184, 208, 212

embolism, xii, 167, 183

embolization, 208

emergency, 9, 124, 154, 155, 157, 158, 159

emotion, 6, 23

encephalitis, 118, 126, 129, 140, 162

encephalomyelitis, vii, x, 109, 110, 117, 119, 120, 121, 122, 126, 132, 139

encephalopathy, 139, 141, 144, 152, 154, 157, 158

encoding, 38, 42, 72

endothelial cells, 82

endothelium, 69

energy, 9, 52, 142

environment, 9, 13

environmental factors, 3

enzyme(s), 71, 75, 84

epidural hematoma, 151

epilepsy, 15, 16, 19, 20, 24, 126, 128

epinephrine, 14

episodic memory, 21

epitopes, 118

ester, 71

estrogen, 207, 209

etiology, 118, 124, 128, 129, 165, 182, 183, 184, 187, 216

eukaryotic, 107

everyday life, 21

evolution, 6, 7, 8, 12, 21, 22, 116

excitability, 15

excitotoxicity, 61, 83

exclusion, 202

excretion, 198, 200, 212, 218

executive function(s), 8, 22

extravasation, 212

eye movement, 6, 30

F

facial nerve, 77

family history, 50
feelings, 3
ferromagnetic, 51, 57
fever, xi, 124, 125, 131
fiber(s), 10, 62, 65, 73, 117, 179, 180, 181, 186, 193
fibrillation, 184
fibrin degradation products, 209
fibrinogen, 199, 205, 206, 209, 210, 212, 213
fibroblasts, 84
filament, 100, 103
filtration, 204, 210
first degree relative, 93
first generation, 51
flexor, 176
fluid, xi, 14, 111, 124, 133, 159, 185
fluorogold, 75
focal segmental glomerulosclerosis, 215
focal seizure, 111
foramen, 60, 182
foramen ovale, 182
forebrain, 6, 13, 14, 15, 155
formation, 6, 14, 71, 72, 100, 102
fragments, 90, 100
Freud, Sigmund, 10, 12, 24
frontal cortex, 8, 14, 91
frontal lobe, 93, 175, 205, 211
functional MRI, 169, 194

G

GABA, 83
gait, 128, 141, 208
gallbladder, 151
gamma frequency, viii, 27, 28, 31, 33, 41
gastrointestinal tract, 147
gastroparesis, 160, 163
GDNF, ix, 60, 64, 65, 66, 68, 72, 73, 77, 78, 79, 85, 86
gender differences, 30, 160, 163
gene expression, 77
gene therapy, 16
gene transfer, 85
general anesthesia, 9

general practitioner, 168
general surgery, 18
genes, x, 3, 8, 63, 90, 96
genetic defect, 92, 93
genetic factors, x, 90, 100
genetic linkage, 105
genetics, 104, 106
genome, 16
gland, 8, 9
glia, 10, 83, 112, 113
glial cells, 63, 91
glioma surgery, 16
glomerulonephritis, 202, 207, 217, 220, 222
glucose, 125, 126, 127, 128
glutamate, 83
glycine, 90
granules, 74
growth, viii, 2, 8, 9, 12, 13, 15, 64, 76, 78
growth factor, viii, 2, 15, 64, 76, 78
growth hormone, 15
Guillain-Barre syndrome, 110, 117, 121, 140

H

headache, xi, 111, 116, 124, 125, 131, 140, 205, 206, 208, 219, 221
health problems, 30
heart disease, 17, 220
heavy drinking, 207, 209, 214
hematocrit, 210
hematoma, 138, 151, 205
hemiparesis, 111, 126, 170, 176, 181, 188, 206, 208
hemiplegia, 193, 205, 207
hemisphere, 6, 169, 211, 212
hemorrhage, xiii, 22, 24, 50, 57, 147, 170, 174, 188, 198, 200, 201, 208, 210, 211, 212, 216, 217, 219, 221
hemorrhagic stroke, 169, 170, 173, 185
hemostasis, 18
hepatic encephalopathy, 144, 152
herbal medicine, 128
herpes simplex, 124, 126, 130, 132
herpes virus, 126

herpes zoster, 144

heterogeneity, 102, 103, 120

high density lipoprotein, 204, 210

hippocampus, 6, 15, 21, 28, 92, 93, 95, 98, 99

histochemistry, 70, 76

histology, 114

HLA, 112, 113, 114

homeostasis, 18

homopolymer polyacetal, 52, 56

hormone, 15, 79, 151, 206, 209

human behavior, 5

human brain, vii, viii, 2, 3, 6, 7, 8, 12, 13, 17, 18, 21, 193, 194

hybrid, xi, 109, 110, 115

hybridization, 70, 86

hydrocephalus, 151, 160, 161, 165

hyperactivity, 148

hypercholesterolemia, 205

hyperlipidemia, xii, 184, 197, 198, 200, 213

hyperplasia, 128, 145, 151

hypertension, xiii, 153, 184, 198, 204, 208, 212, 213, 214

hypertrophy, 134

hypotension, 140, 141, 146, 148, 183, 189

hypothalamus, 7, 8, 9, 12, 146

I

idiopathic, 144, 151, 161, 220

image(s), 21, 70, 174, 178, 185, 194, 211

immune activation, 120

immune response, 9

immunoglobulin, 114, 152, 155

immunohistochemistry, 75, 91, 105, 111, 114

immunoreactivity, 69, 73, 99, 101, 102

implants, 15, 20

impulses, 19

in situ hybridization, 70

in vitro, 65

in vivo, 65, 78, 84, 146

induction, 70, 80

infarction, xii, 129, 153, 158, 167, 168, 169, 170, 173, 174, 175, 176, 177, 178, 179, 180, 181, 182, 185, 186, 187, 188, 189, 190, 191, 192, 193, 194, 195, 202, 218, 219

infection, 84, 116, 141

inflammation, 55, 72, 73, 76, 84, 110, 114, 119

inflammatory demyelination, 117, 118, 119, 121

informed consent, 30, 34

inhibition, x, 12, 13, 60, 72, 84, 141, 142

inhibitor, 70, 71, 72, 73, 84, 85, 86, 144, 148, 152, 162

injury(ies), viii, ix, xi, 2, 14, 16, 18, 59, 60, 61, 62, 63, 65, 66, 67, 68, 69, 70, 72, 73, 74, 75, 76, 77, 78, 80, 81, 82, 83, 85, 87, 124, 130, 131, 150, 192

intelligence, 24, 165

interaction effect(s), 36, 37, 38, 40

interstitial nephritis, 213

intracerebral hemorrhage, xii, 174, 188, 197, 200, 201, 208, 210, 211, 216, 217, 221

intracranial aneurysm, 50, 56, 57

intracranial pressure, 14, 199

intraocular, 79

introspection, 5

inversion recovery, 178, 185

ipsilateral, 65, 105, 169

iron, 214

irritability, 12

irritable bowel syndrome, 165

ischemia, viii, 2, 16, 24, 185

K

kidney, 205

L

labeling, 75

lamination, 7

laparoscopy, 58

L-arginine, 69, 71

latency, 35, 38, 41, 43

lateral sclerosis, 91

laws, 3, 4, 21, 22

LDL, 144, 204, 210

leakage, ix, 50, 55, 71, 75

learning, 6, 7, 12, 13, 14, 15, 18

lesions, vii, xii, 12, 15, 18, 20, 82, 91, 96, 102, 104, 110, 111, 113, 114, 116, 117, 118, 119, 120, 128, 129, 140, 149, 152, 158, 159, 167, 169, 170, 175, 180, 181, 182, 184, 187, 188, 191, 204, 214

limb weakness, 117, 183, 190

limbic system, 8, 11, 12, 14

linguistics, 5

lobectomy, 16, 20

local anesthesia, 11

localization, 69, 79, 86, 189, 190

locus, 14

lumbar radiculopathy, 192

lumbar spine, 140

lupus, 144, 206, 209, 215, 219

lupus anticoagulant, 206, 209, 219

lupus erythematosus, 219

lymphocytes, 110, 111, 112, 113, 114, 116

lysis, 207

M

macrophages, 73, 111, 114

magnetic resonance imaging (MRI), ix, 49, 53, 54, 110, 111, 113, 116, 117, 125, 126, 128, 129, 130, 139, 141, 160, 163, 168, 169, 173, 175, 180, 181, 184, 187, 189, 190, 192, 193, 194, 198, 202, 205, 206, 208, 210, 211, 212

major depression, 159

mammalian brain, 12, 79

manipulation, vii, x, 60, 71, 86

mapping, 23, 161, 194

measurement(s), 74, 75, 102, 140, 154, 204

mechanical properties, 55

medication, 30, 203

medulla, 6, 14, 181

medulla oblongata, 181

melatonin, 9

mellitus, 184, 202, 203, 204, 206, 210, 221

membranes, 62, 63

membranoproliferative glomerulonephritis, 202, 217

membranous glomerulonephritis, 207, 220

membranous nephropathy, 213, 215

memory, vii, ix, 6, 8, 14, 15, 21, 28, 29, 31, 33, 34, 37, 42, 43, 44

memory capacity, vii, 28

memory performance, 42, 43

memory processes, 28

memory retrieval, 34

meninges, 60, 114

meningioma, 16

meningitis, vii, xi, 123, 124, 125, 127, 128, 129, 130, 131, 133, 153, 154

mental disorder, 17

mental energy, 9

mental life, 19

mental processes, viii, 2, 6, 11, 43

mental state(s), 7

mesangial proliferative glomerulonephritis, 207

mesencephalon, 6

meta-analysis, 16, 57, 221

metabolism, 8, 14, 218

methylene blue, 74

methylprednisolone, 117

microdialysis, 146

microscopy, 58, 117

midbrain, 6, 14, 78, 181

minimal change disease, 215, 217

minimal change nephrotic syndrome, 218, 219, 220

Ministry of Education, 138

mitochondria, 62, 71, 77

model system, 104

modules, 40, 99

molecular biology, 5

molecular pathology, 99

molecular weight, 199

molecules, 64

MOMP, 71

monoclonal antibody, 86, 93

morbidity, 17, 50, 57, 204

morphology, ix, 50, 59, 62, 74, 82

morphometric, 102
mortality, 17, 21, 50
Motoneuron Death, 59, 62
motor behavior, 6
motor control, 175
motor fiber, 186, 193
motor neuron disease, 91, 103, 105, 107
motor neurons, 76, 87
movement disorders, 15
mRNA, 65, 70, 75, 79, 80, 86
multiple myeloma, 214
multiple sclerosis, 15, 110, 115, 119, 120, 121, 129, 140
muscle atrophy, 129
muscle contraction, 160, 163
muscles, 61, 68, 73, 82, 176, 184
mutant, 100
mutation(s), x, 89, 92, 93, 94, 96, 98, 100, 101, 103, 104, 105, 206, 209, 218, 220
myelin, xi, 3, 13, 111, 124, 125, 126
myelin basic protein (MBP), xi, 111, 124, 125, 126, 128

N

narcolepsy, 15
National Academy of Sciences, 44, 46, 76, 78, 80, 84
necrosis, 62, 69
neocortex, 6, 7, 9, 12, 14, 15
nephritic syndrome, 201, 207
nephritis, 207, 213, 215
nephropathy, xiii, 198, 202, 203, 204, 210, 212, 213, 215, 217, 221, 222
nephrotic syndrome, vii, xii, 197, 199, 202, 203, 204, 206, 207, 209, 210, 214, 216, 217, 218, 219, 220, 221, 222
nerve, x, xi, xii, 10, 20, 60, 61, 62, 64, 67, 68, 73, 75, 76, 77, 78, 80, 81, 82, 83, 86, 110, 117, 118, 125, 127, 129, 134, 149, 168, 176, 177, 184, 192, 208
nerve fibers, 10
nerve growth factor (NGF), 64, 76, 78
nervous system, viii, 2, 4, 6, 8, 9, 10, 15, 16, 18, 65, 78, 119, 124, 125, 129, 132, 169, 184
neural network(s), vii, 8, 14, 21
neuritis, 10, 101, 117, 121
neurodegeneration, x, 73, 83, 89, 90, 119
neurodegenerative diseases, 66, 72, 104
neurodegenerative disorders, 20, 102
neurofibrillary tangles, 102
neurogenic bladder, 143
neuroimaging, 188
neuroleptic malignant syndrome, 148
neurological disease, viii, 2, 15, 16
neurologist, 12, 18
neuroma, 74
neuromyelitis optica, 129
neuronal apoptosis, 62, 85
neurons, vii, 1, 3, 6, 7, 9, 10, 11, 13, 14, 20, 64, 65, 66, 69, 71, 72, 74, 75, 76, 77, 78, 79, 80, 81, 82, 85, 86, 87, 90, 91, 105, 143
neuropathy, 15, 117, 121, 122, 129, 142, 144, 147, 150, 157, 168, 184, 190
neurophysiology, 5, 80
neuroprotection, 66, 73
neurosarcoidosis, 142
neuroscience, vii, viii, 2, 5, 10, 16, 17, 23, 24, 76, 77, 80, 82, 83, 84, 85, 160
neurosurgery, vii, ix, 4, 15, 16, 17, 18, 19, 49, 50, 51, 55, 56, 57, 77, 81
neurotoxicity, 85
neurotransmitter(s), 6, 14, 75, 83
neurotrophic factor(s), ix, 60, 61, 62, 64, 65, 66, 67, 73, 76, 86
nigrostriatal, 142
Nissl Staining, 74
nitric oxide (NO), x, 60, 63, 69, 70, 73, 79, 80, 83, 84
nitric oxide synthase (NOS), x, 60, 61, 63, 66, 69, 70, 73, 76, 77, 79, 80, 83, 84, 86
nocturia, 154
nodules, 120
non-adrenergic non-cholinergic, 83
norepinephrine, 14
normal pressure hydrocephalus, 151, 161

nuclear membrane, 63

nuclei, 6, 10, 12, 14, 15, 20, 113

nucleic acid, 74

nucleus, 10, 14, 15, 20, 63, 74, 77, 90, 100, 144, 146, 180

nystagmus, 162

O

obsessive-compulsive disorder (OCD), 19

obstruction, 124, 141, 142, 147, 148, 149, 155, 157, 159

occipital cortex, 116

occipital lobe, 211

occlusion, xii, 177, 183, 184, 198, 201, 202, 203, 206, 212

oligodendrocytes, 62, 111, 113, 114

oligodendroglia, 114

operations, 11, 19

optic neuritis, 117, 121

optimism, 20, 21

organ(s), 2, 3, 143, 147, 152, 154

orthostatic hypotension, 141

oscillatory activity, 28, 34

ossification, 138

otitis media, 209, 219

overlap, 115, 180, 187

oxidative damage, 77

oxidative stress, 76

oxygen consumption, 69

oxyhemoglobin, 154

P

p53, 77

p75, 64, 66, 77

pain, 14, 18, 20, 124, 125, 154

papilledema, 207

paralysis, 147, 176, 177, 192

parasympathetic nervous system, 9

parathyroid hormone, 79

parenchyma, 114, 116

paresis, xi, 147, 167, 169, 173, 174, 176, 178, 179, 181, 183, 185, 187, 189, 190, 191, 193

parietal lobe, 91, 111, 175, 182, 188

parkinsonism, 91, 104

paroxetine, 150, 151

partial thromboplastin time, 210

participants, viii, 27, 28, 29, 30, 34, 36, 38, 42, 43

pathogenesis, 102, 104, 110, 120, 200, 213

pathology, vii, xi, 92, 99, 101, 102, 103, 104, 105, 106, 114, 118, 124, 125, 129, 132, 182, 213, 221

pathophysiological, 16, 162

pathophysiology, 18, 134, 218

pathway(s), 18, 62, 71, 74, 77, 79, 83, 85, 133, 147, 86

pelvic floor, 160, 163

peptide(s), viii, 2, 15, 72, 79, 105

peripheral nerve graft, x, 60, 67, 68, 80, 81

peripheral nervous system, 124, 132, 184

peripheral neuropathy, 117, 121, 168, 190

personality, 207

PGN, 202, 217

phenotype(s), 103, 110, 118

phosphorylation, x, 90, 92, 99, 101

physical properties, 52

physicians, 4, 187

physiology, 10, 18, 135

phytotherapy, 145

PI3K, 63

pia mater, 92, 95, 96, 97, 100

pineal gland, 9

pituitary gland, 8

placebo, 140, 153

plaque, 102, 111

plasminogen, 199

plastic clips, vii, ix, 49, 50, 51, 52, 55, 56, 57

plasticity, viii, 2, 13, 15, 66

platelet aggregation, 199

platelet count, 210

pleural effusion, 208

plexus, ix, 59, 60, 76, 78, 81, 82, 125

pneumonia, 211

polymer, 52, 57
polymerase chain reaction, 119
polyuria, 143, 146
pons, 6, 14, 181, 194
positive correlation, 31, 33
postural hypotension, 148
preeclampsia, 207, 209, 219
prefrontal cortex, 8, 21
pregnancy, 152, 155, 199
prevention, 15, 168
primary amyloidosis, 202
primate, 7, 12, 23, 87
prognosis, vii, x, xii, 16, 109, 168, 185, 186, 187, 189
progressive supranuclear palsy, 96, 138
prolapse, 144
prosthesis, 15
protection, viii, 2, 16, 60, 62, 63, 66
protective genes, 63
proteins, 66, 90
proteinuria, xii, 197, 198, 200
proteolysis, 72
proteolytic enzyme, 71
prothrombin, 210, 218
prothrombin time, 210
protoplasm, 10
psychiatric disorder(s), 15, 19
psychiatric patients, 19
psychosis, 24
psychosurgery, 18, 19
psychotic symptoms, 149
PTP-σ, 72
PTT, 210
public awareness, 15
puerperium, 199
pyramidal cells, 7
pyridostigmine, 148

Q

quality of life, 20

R

radiation, 138
radiculopathy, 192
reaction time, viii, 27, 34, 36, 37, 42, 43
reactions, 55
reactivity, 69
reading, 92
reasoning, 16, 28
recall, 21
receptor(s), 14, 15, 64, 71, 72, 76, 79, 80, 85, 86, 152
recognition, 38, 90, 118
reconstruction, 73
recovery, 68, 72, 76, 81, 82, 85, 86, 110, 115, 176, 177, 178, 185, 186, 191
recurrence, 110, 111, 185
reflexes, 125, 127, 128, 129, 176, 184
regenerate, x, 60, 67, 68
regeneration, viii, x, 2, 16, 60, 63, 65, 68, 72, 73, 75, 76, 78, 81, 82, 85, 86
registries, 215
registry, 188, 191, 215, 222
rehabilitation, 208, 212
rejection, 13
relapses, x, 109, 115
relatives, 93
relaxation, 69, 82, 141, 142
remission, 207, 214
remyelination, 117
repair, ix, 18, 60, 67, 76, 80, 81
resources, 40, 41, 43
respiration, 7, 83
response, x, 35, 38, 71, 89, 90, 117, 145
responsiveness, 9, 14
restoration, ix, 59
reticular activating system (RAS), 14
reticulum, 11, 74, 100
retrograde labeling, 75
rheumatoid arthritis, 207
rhythm, 146, 150
ribosomal RNA, 74
risk factors, xii, xiii, 168, 186, 198, 199, 200, 213, 214, 216, 218

risk(s), xii, xiii, 4, 13, 17, 106, 116, 117,
143, 159, 168, 186, 198, 199, 200, 212,
213, 214, 215, 216, 217, 218, 220, 221
RNA, 74, 90, 100
root(s), vii, ix, x, 59, 60, 61, 62, 63, 65, 67,
68, 69, 70, 72, 73, 75, 76, 77, 78, 79, 80,
81, 82, 83, 84, 85, 86, 87, 117, 127, 134,
150
rubella, 153

S

saddle anesthesia, 124
safety, 13
SAR, 59
Sarajevo, 1, 24, 25, 49
sarcoidosis, 143
scar tissue, 73
sclerosis, 15, 91, 102, 110, 115, 119, 120,
121, 129, 140, 213, 216
secretion, 100, 149, 151
segmental glomerulosclerosis, 215, 222
seizure, 20, 176, 207
selective attention, 37, 43
self-reflection, 8
sensation seeking, 13
sensation(s), 7, 13, 127, 144, 148, 149, 157,
179
senses, 161
sensitivity, 187
sensory symptoms, 168, 169, 175, 177, 181
sepsis, 115
serotonin, 14
serum, 128, 200, 213
serum albumin, 200
sham, viii, 27, 34, 35, 36, 37, 38, 39, 40, 41,
42, 43
shape, 8, 96
shock, 126, 127, 128
short term memory (STM), viii, 27, 28, 29,
32, 33, 41, 42, 43
showing, xi, 33, 63, 66, 70, 72, 109, 112,
113, 115, 118, 174, 178, 193, 200, 211
shunt surgery, 161
signal peptide, 105

signalling, 79, 85
signals, 3, 71, 211
signs, xi, xii, 113, 120, 124, 125, 126, 128,
129, 131, 168, 174, 180, 184, 208, 211,
212
silver, 10, 51
sinuses, 206, 218
skeletal muscle, 79, 81, 82
skin, 124, 125, 149
sleep apnea, 157
smoking, 213
smooth muscle, 69, 82
social activities, 111
spatial information, 28
species, 13, 69
speech, 169, 208
sphincter, 124, 125, 126, 127, 128, 139,
142, 143, 145, 146, 149, 150, 158, 159
spinal bifida, 132, 143
spinal cord, x, 14, 60, 62, 65, 66, 67, 68, 70,
72, 73, 74, 76, 78, 81, 82, 85, 86, 126,
127, 128, 129, 133, 143, 144, 147, 148,
151
spinal cord injury, 72, 85
spinal cord tumor, 144
spinal root avulsion, ix, 59, 60, 75, 78
spine, 19, 138, 140, 161
spinocerebellar degeneration, 140
sprouting, 87
SSS, 206, 207
standard deviation, 31
state(s), xiii, 5, 9, 7, 11, 14, 198, 200, 205,
213, 215
stem cells, viii, 2, 15
stenosis, 177, 182, 184, 186, 210, 211, 212
stimulation, viii, 15, 18, 19, 20, 25, 27, 29,
33, 34, 35, 36, 37, 38, 39, 40, 42, 68, 80,
81, 140, 141, 142, 144, 149, 152, 160,
163, 190
stimulus, 34, 35, 38
storage, 9, 28, 33, 126, 143, 146
stress, 17, 55, 76, 140
striatum, 14, 152
stroke, vii, viii, xi, xii, 2, 15, 16, 20, 139,
146, 151, 154, 156, 159, 167, 168, 169,

170, 171, 173, 174, 177, 182, 183, 185, 186, 187, 188, 189, 190, 191, 192, 195, 197, 198, 200, 201, 202, 203, 204, 210, 212, 213, 214, 215, 217, 218, 219

subacute, 129, 142

subarachnoid hemorrhage, 22, 50, 201

subarachnoidal, 212

subgroups, xi, 109, 118, 122, 124, 182

substrate(s), viii, 2, 17, 69, 71

sulfate, 72, 85

suppression, 76, 158, 159

survival, vii, ix, 21, 59, 63, 64, 65, 66, 67, 68, 70, 71, 72, 73, 76, 78, 79, 80, 81, 82, 84, 85, 86, 105

survival rate, 67, 68, 73

sympathetic nervous system, 9, 78

symptoms, vii, xi, 19, 111, 117, 124, 125, 131, 149, 150, 151, 154, 161, 165, 168, 170, 175, 177, 181, 199, 206

synaptic plasticity, 66

synchronization, 42

synergistic effect, 85

syringomyelia, 139

systemic lupus erythematosus (SLE), 207, 209, 215, 219

T

T cell(s), 112, 113, 114, 116, 118, 119

T lymphocytes, 111, 113

tACS, viii, 27, 28, 29, 33, 34, 35, 36, 37, 38, 39, 40, 41, 42, 43, 44, 46, 47

tangles, 102

target, 20, 68, 72, 73

task difficulty, 41

tau, 92, 99, 101, 103, 104, 105

techniques, vii, 2, 5, 18, 20, 74

technology(ies), 3, 57

telencephalon, 6, 11

temporal lobe, x, 7, 16, 20, 90, 93, 100, 208, 211, 212

temporal lobe epilepsy, 20

tendon, 129, 176

tension, ix, 50, 55, 56

thalamus, 6, 7, 15, 20

therapy, viii, xi, 2, 15, 16, 18, 20, 124, 130, 131, 150, 162, 207, 209

theta frequency, viii, 27, 28, 33, 41, 43

thoracic spine fracture, 161

thrombosis, xii, 182, 197, 198, 199, 200, 201, 206, 207, 209, 212, 214, 216, 217, 218, 219, 220, 221

tics, 45, 48

tissue, x, 10, 60, 73, 79, 90, 95, 96, 97, 99, 102, 115

titanium, ix, 49, 50, 51, 52, 53, 55

titanium clips, ix, 49

togo, 105

tonic, 35

total cholesterol, 201, 202, 206, 210

toxicity, 70, 72

traditional views, 175

training, 18, 142

traits, 13

trajectory, 68

transcription, 63, 90

transection, 78, 83

transient ischemic attack, 190, 201

transplantation, viii, 2, 16, 18, 20, 67, 82

trauma, 61, 67, 72

traumatic brain injury, 18, 72, 85

treatment, vii, viii, 2, 3, 4, 15, 16, 18, 19, 20, 24, 50, 56, 65, 66, 70, 72, 73, 80, 110, 134, 140, 144, 145, 149, 150, 151, 153, 165, 199, 207, 208, 212, 214, 217, 220

Trk, 64

TrkB, 66, 80

tumor(s), 3, 12, 16, 18, 65, 119, 138, 144, 169

type 2 diabetes, 221

tyrosine, 64, 72

U

ubiquitin, x, 90, 92, 94, 96, 103, 104, 105, 107

ultrasonography, 210, 212

ultrasound, 204

ultrastructure, 117

underlying mechanisms, 114
United States (USA), viii, 2, 15, 19, 20, 31,
    35, 76, 78, 80, 84, 93, 94, 101, 221
universe, vii, viii, 1, 2, 4, 8, 13, 21, 22
upper respiratory tract, 116
urinary dysfunction, 142, 150, 151, 152,
    155, 159, 160, 163
urinary retention, xi, 123, 124, 125, 126,
    127, 128, 129, 130, 131, 132, 133, 143,
    144, 146, 147, 148, 149, 154, 157
urinary tract, 126, 132, 138, 141, 144, 146,
    147, 151, 154, 159, 161, 164
urine, 150, 199, 200, 201

V

vaccinations, 118
vascular dementia, 211
vasodilation, 149
vasopressin, 146, 150
vasospasm, 24
vein, 207, 212, 218, 220, 221
ventral root (VR) re-implantation, 67
vertebral artery, 184
vertebrates, 6, 7
vertigo, 156, 158, 159, 212

vessels, 11, 18, 69, 203, 215
vision, 208
visual field, 169
visual processing, viii, 28, 37, 42, 43
vomiting, 149, 205, 208, 214
VR re-implantation, 67
vulnerability, xiii, 198, 214

W

waking, 143
walking, 6
weakness, xii, 117, 129, 152, 167, 168, 169,
    175, 176, 177, 179, 180, 181, 183, 184,
    185, 186, 187, 188, 189, 190, 191, 192
white matter, 6, 19, 95, 111, 112, 113, 114,
    116, 120, 121, 126, 129, 158, 161, 162,
    175, 176, 182
working memory, vii, viii, 8, 27, 28, 29, 31,
    33, 34, 37, 42, 44

Z

zinc, 83